Bridges Between Palm Beach County & Miami, Florida

Bridges Between Palm Beach County & Miami, Florida

By Harry McQuire + Alex McQuire

Humanics Publishing Group
Atlanta, Lake Worth

Humanics Publishing Group

Bridges Between Palm Beach County & Miami, Florida

A Humanics Publishing Group Publication

First Edition

Brumby Holding, Inc.
PO Box 1608
12 S. Dixie Highway, Ste. 203
Lake Worth, FL 33460
USA

www.humanicspub.com

Printed in the United States of America

ISBN # 0-89334-466-4

This book is dedicated to our family.

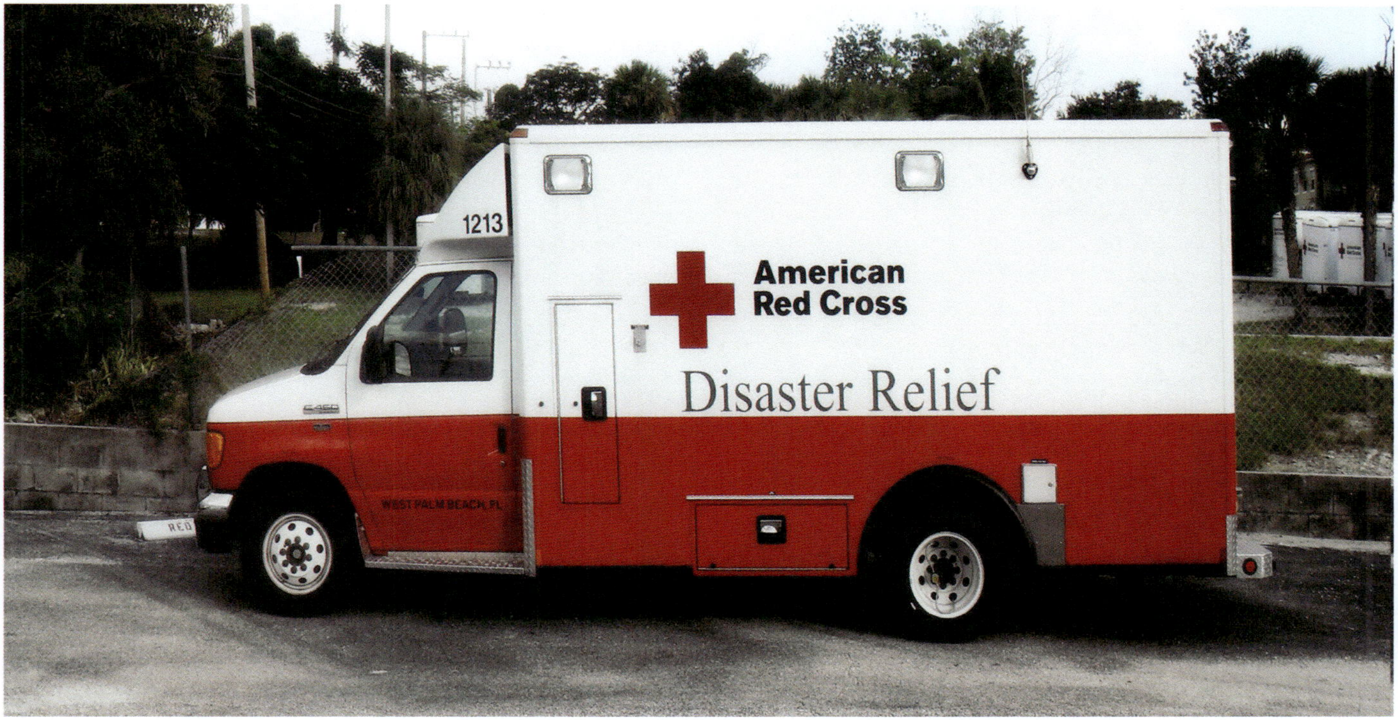

Special Acknowledgement

We would like to acknowledge the American Red Cross, a nonprofit humanitarian organization that responds to our communities needs. They not only provide emergency assistance and disaster relief but they offer training and educational programs on health and safety. We thank the Red Cross and the caring people who support their work.

The Development of Bridges

Bridges are structures built to span a valley, road, body of water, or other physical obstacle, to provide access over the impediment. They are designed for trains, pedestrian or road traffic, and pipelines for water transport. They also allow passage to barges, ships and boats on the waterways.

Their designs vary depending on function and the nature of the terrain where they are constructed. Going back in time, the first bridges were made by nature. They could be as simple as a log fallen across a stream. However, mans' first bridges were probably made from spans of wooden logs or planks and eventually stones; using a simple support and crossbeam arrangement.

Some early Americans used trees or bamboo poles to cross small caverns or wells to get from one place to another. The greatest bridge builders of antiquity were the ancient Romans. The Romans built arch bridges and aqueducts that could stand in conditions that would destroy earlier designs. Some stand today. An example is the Alcantara Bridge, built over the river Tagus, in Spain. The Romans also used cement, which reduced the variation of strength found in natural stone. One type of cement, called pozzolana, consisted of water, lime, sand, and volcanic rock. Brick and mortar bridges were built after the Roman era, as the technology for cement was lost then later rediscovered.

There are several types of bridges, beam bridges are horizontal beams supported at each end by piers. The earliest beam bridges were simple logs that sat across streams. In modem times, beam bridges are large box steel girder bridges. Weight on top of the beam pushes straight down on the piers at either end of the bridge.

Cantilever bridges are built using cantilevers-horizontal beams that are supported on only one end. Most cantilever bridges use two cantilever arms extending from opposite sides of the obstacle to be crossed and meeting at the center. The largest cantilever bridge is the 549-meter (1,800 ft) Quebec Bridge in Quebec, Canada.

Arch bridges are arch-shaped and have abutments at each end. The earliest known arch bridges were built by the Greeks and include the Arkadiko Bridge. The weight of the bridge is thrust into the abutments at either side. Dubai in the United Arab Emirates is currently building the Sheikh Rashid bin Saeed Crossing which is scheduled for completion in 2012. When completed, it will be the largest arch bridge in the world.

Suspension bridges are suspended from cables. The earliest suspension bridges were made of ropes or vines covered with pieces of bamboo. In modem bridges, the cables hang from towers that are attached to caissons or cofferdams. The caissons or cofferdams are implanted deep into the floor of a lake or river. The longest suspension bridge in the world is the 12,826 feet Akashi Kaikyo Bridge in Japan.

Cable-stayed bridges, like suspension bridges are held up by cables. However, in a cable-stayed bridge, less cable is required and the towers holding the cables are proportionately shorter. The first known cable-stayed bridge was designed in 1784 by C.T. Loescher. The longest cable-stayed bridge is the Sutong Bridge over the Yangtze River in China.

Truss bridges are composed of connected elements. They have a solid deck and a lattice of pin-jointed or gusset-joined girders for the sides. Early truss bridges were made of wood, and later of wood with iron tensile rods, but modem truss bridges are made completely of metals such as wrought iron and steel or sometimes of reinforced concrete. The Quebec Bridge, mentioned earlier as a cantilever bridge, is also the world's longest truss bridge.

An aqueduct is a bridge that carries water, resembling a viaduct, which is a bridge that connects points of equal height. A road-rail bridge carries both road and rail traffic. To create a beautiful image, some bridges are built much taller than necessary. This type, often found in east-Asian style gardens, is called a Moon bridge, evoking a rising full moon. Other garden bridges may cross only a dry bed of stream washed pebbles, intended only to convey an impression of a stream. Often in palaces a bridge will be built over an artificial waterway as symbolic of a passage to an important place or state of mind.

The Folklore of the Development of I-95

We've been told that I-95 started out as a trail in the eighteen hundreds by postmen who left Miami and walked north to deliver the mail. This simple trail soon developed into a well worn dirt road. Over time, the road kept evolving into what we have today, a large fast moving artery of cars speeding from place to place like schools of fish.

As we hear, the future will shortly be car guided by electric inputs to stay in lane at fixed speeds, not so exciting as I-95 today.

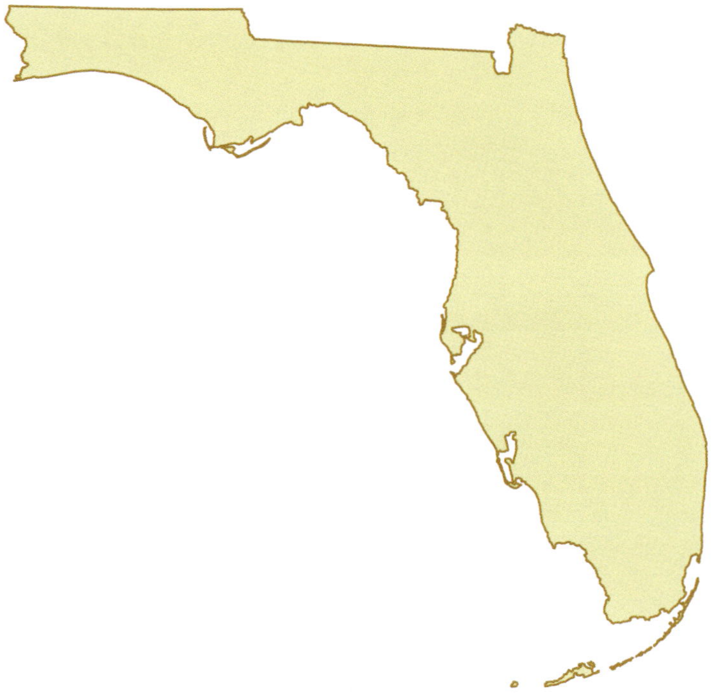

Preface

A short time ago my grandson Alex and I were driving from Miami to North Florida. We looked up while passing under a bridge and Alex pointed out that the bridge abutment above our heads was in different colors. We found that there are at least eight different colors with different narrations in tone. We discussed our interest in bridges, and as I drove, Alex explained what he learned about their structure and how I-95 has been a vital link between North and South Florida.

Alex suggested I should take pictures of the bridges, ramps and roads and put them into a book so that everyone could see for themselves; without taking their eyes off the road.

Enough said, take a look at what Alex and I discovered together!

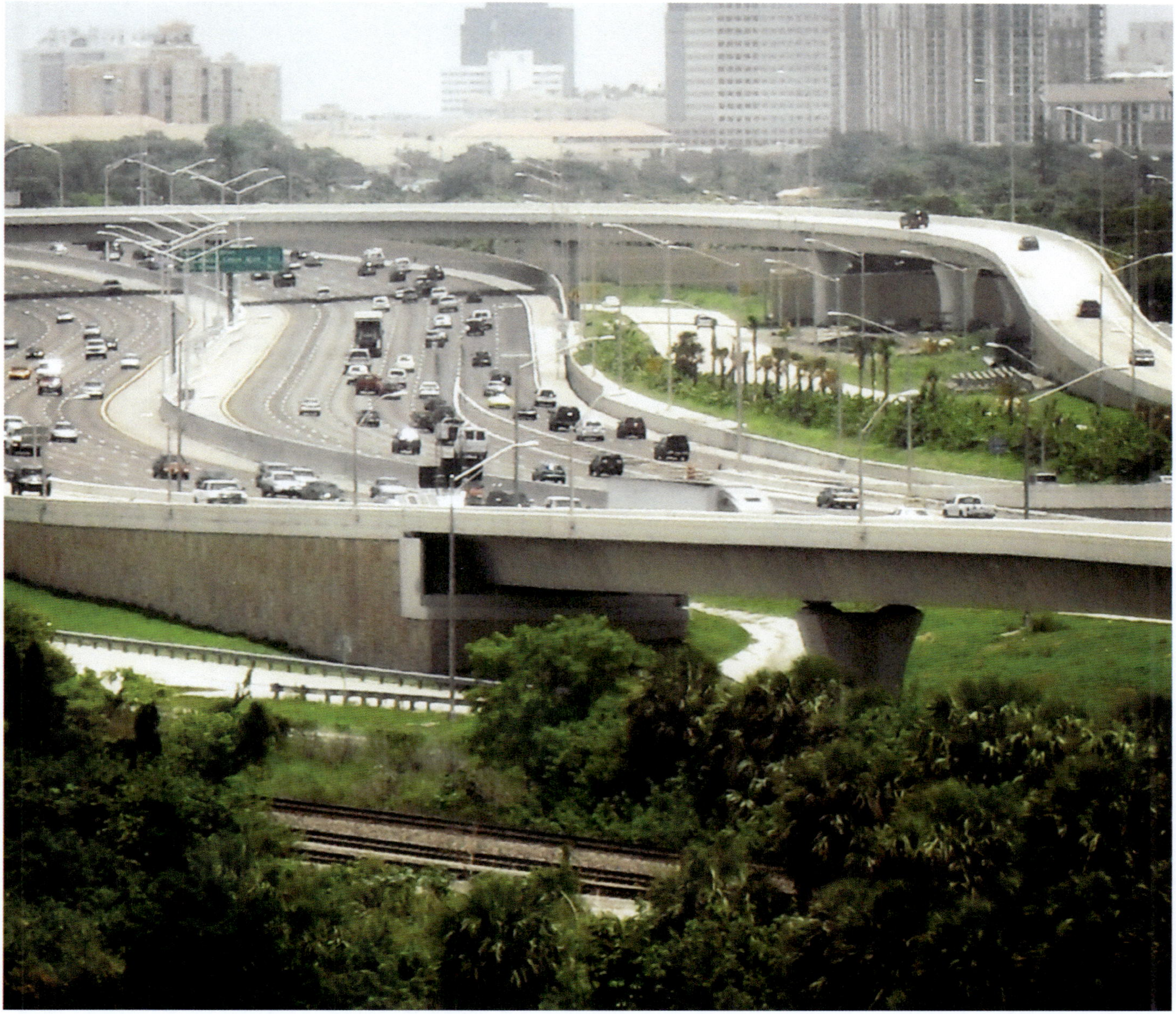

No matter the traffic on the road of life you will eventually make it to the end.

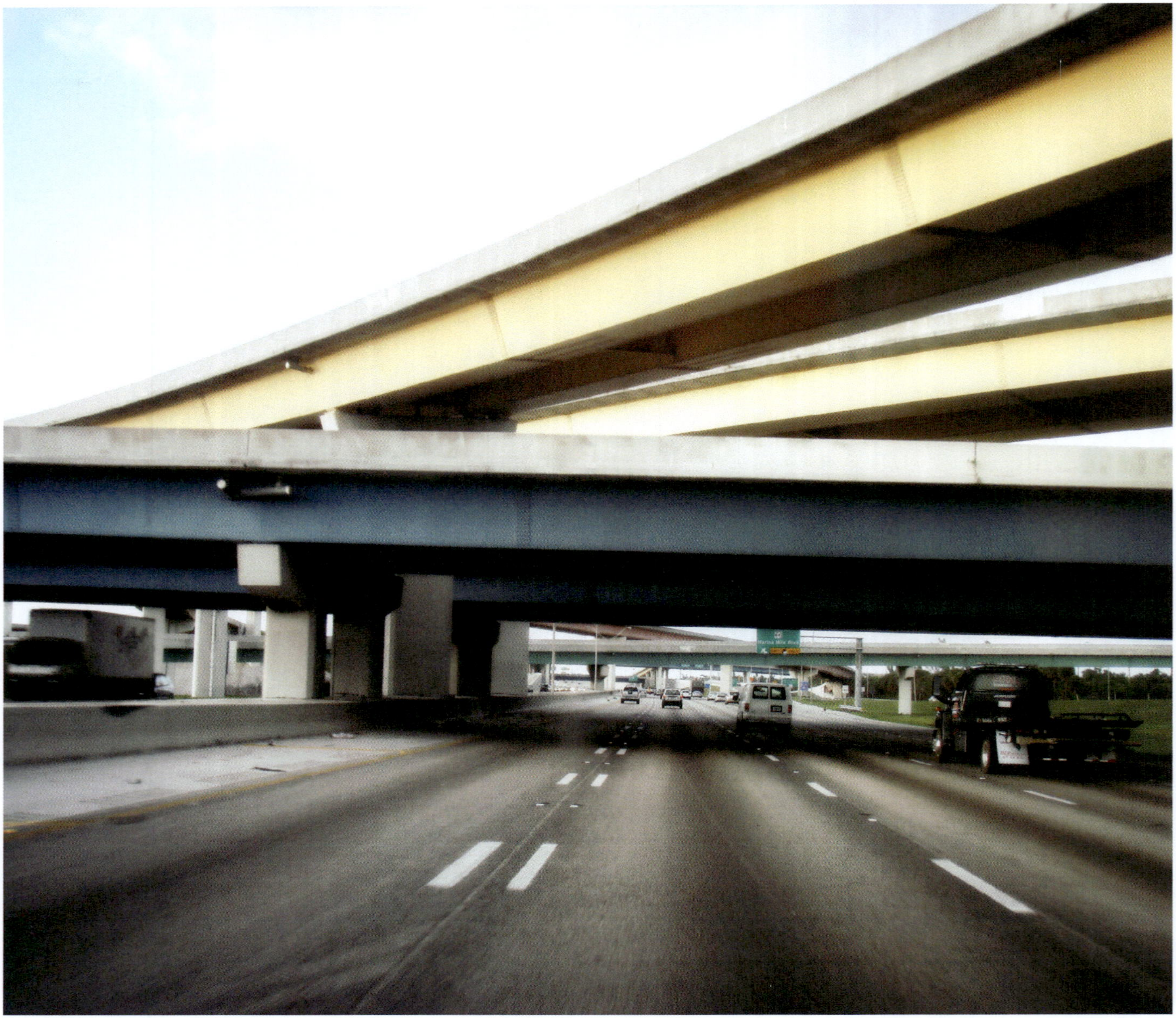

Joy is always down the road.

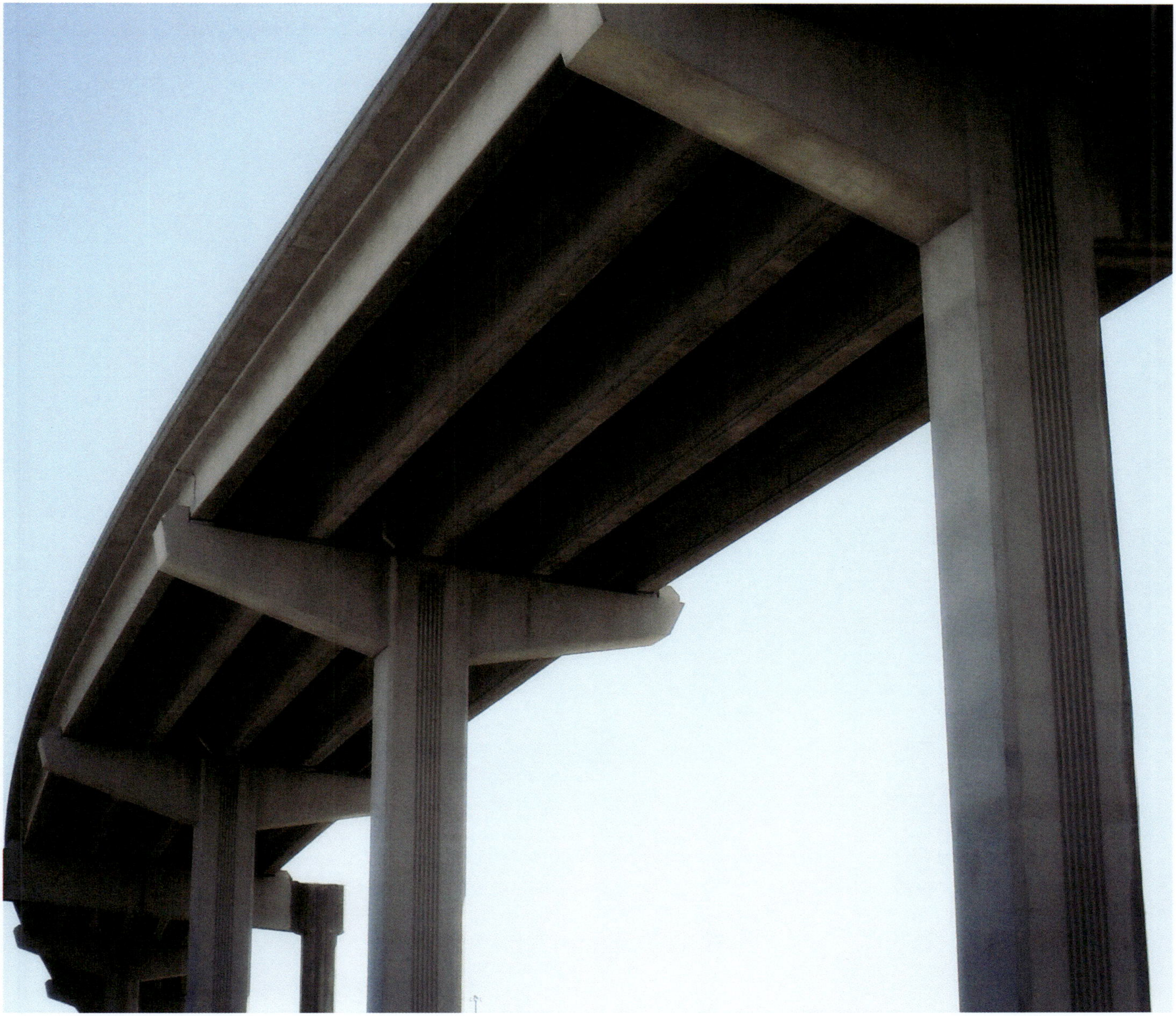

The bridge has a higher goal.

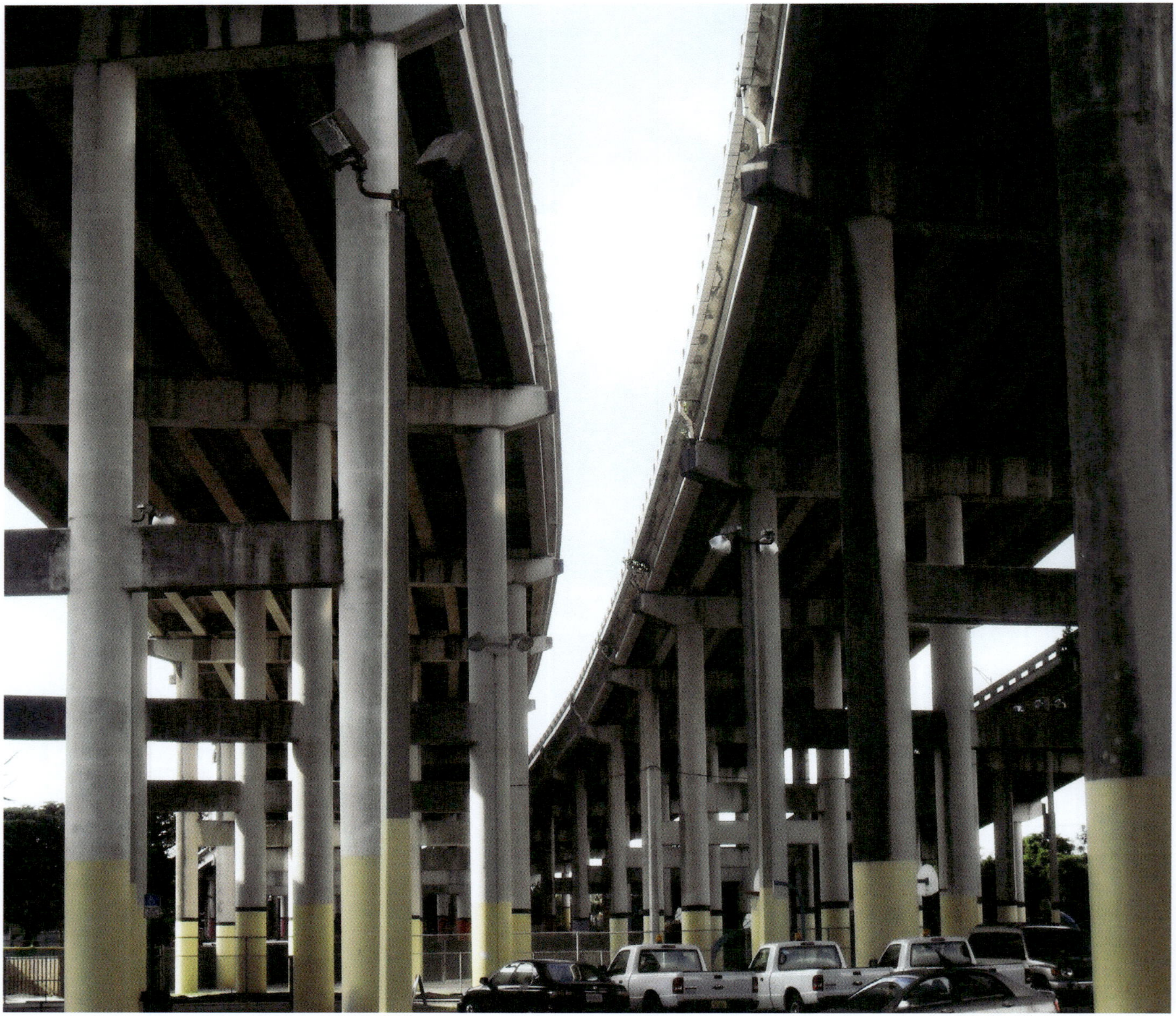

The foundation is every thing.

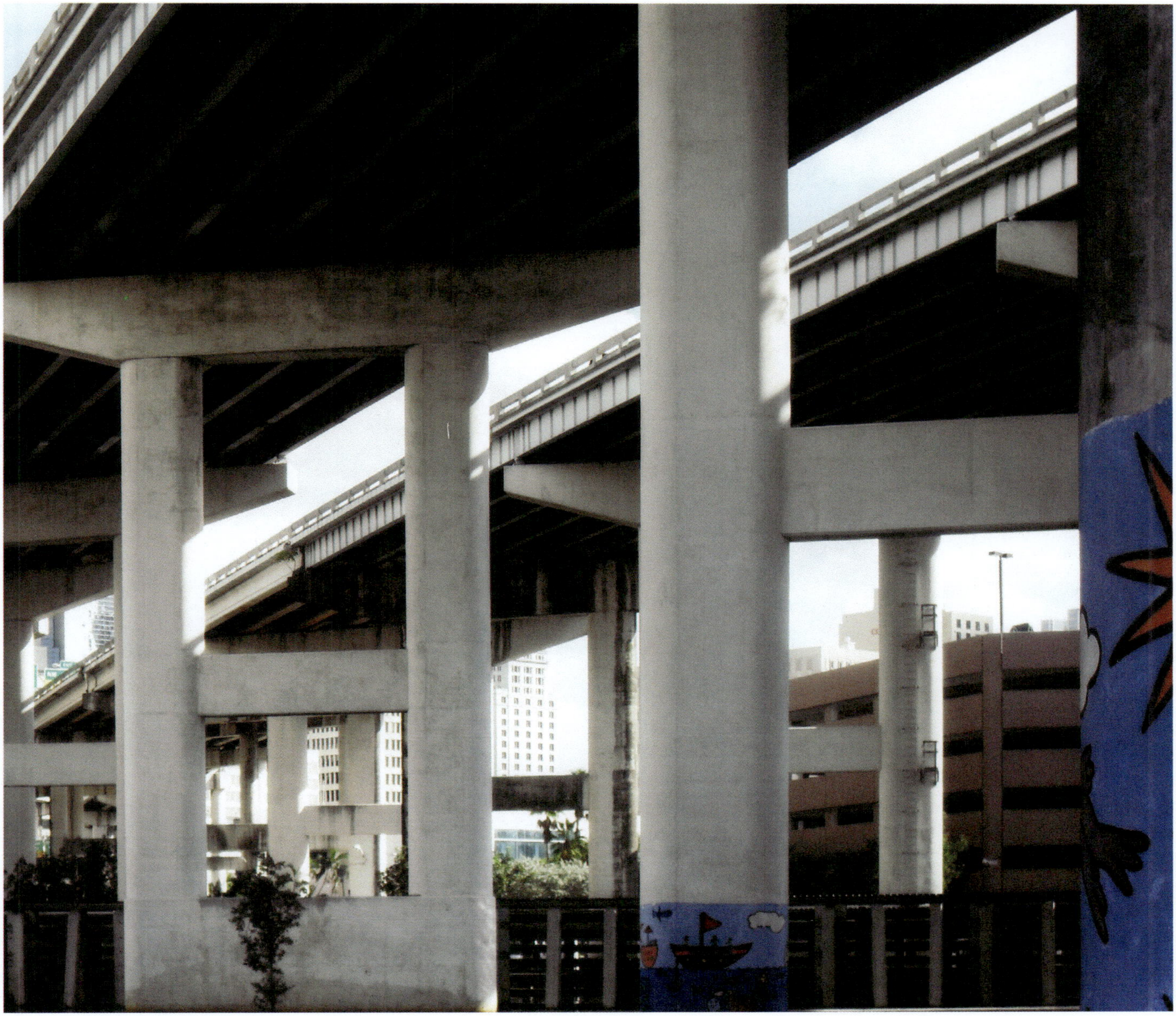

The road needs a bridge for the uncertain continuation of life.

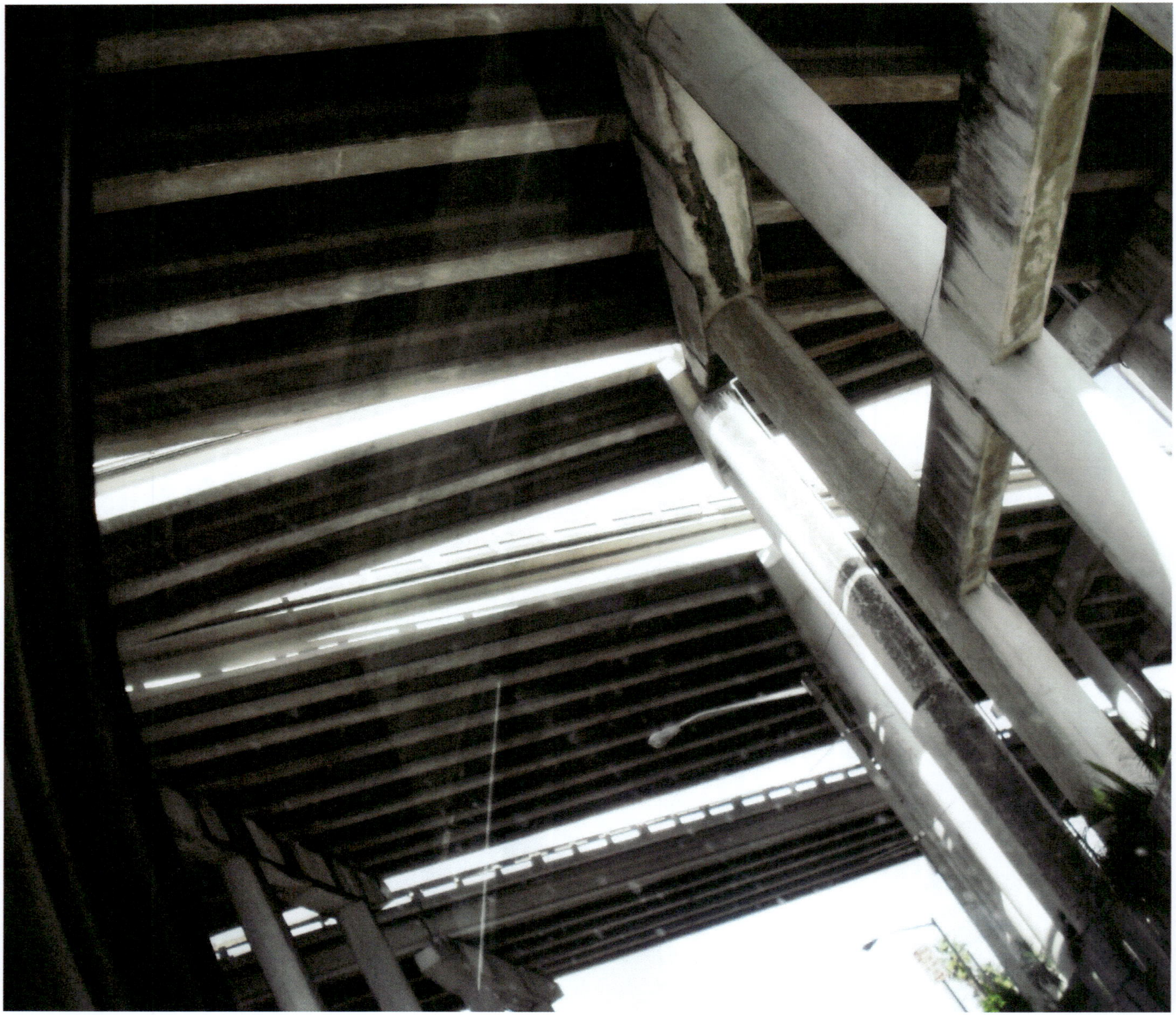

Bridge yourself to the new.

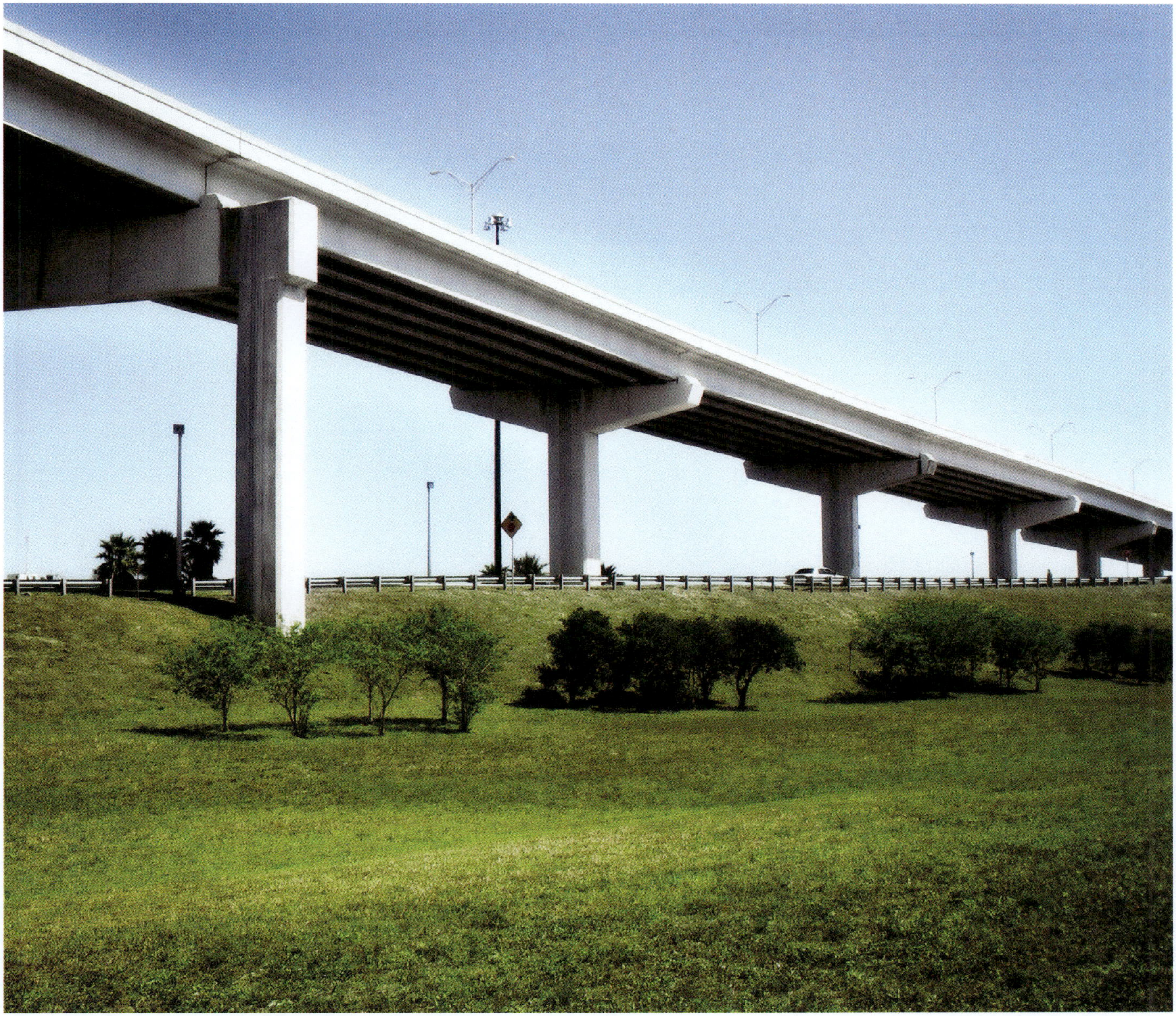

The bridge leads beyond the cosmos.

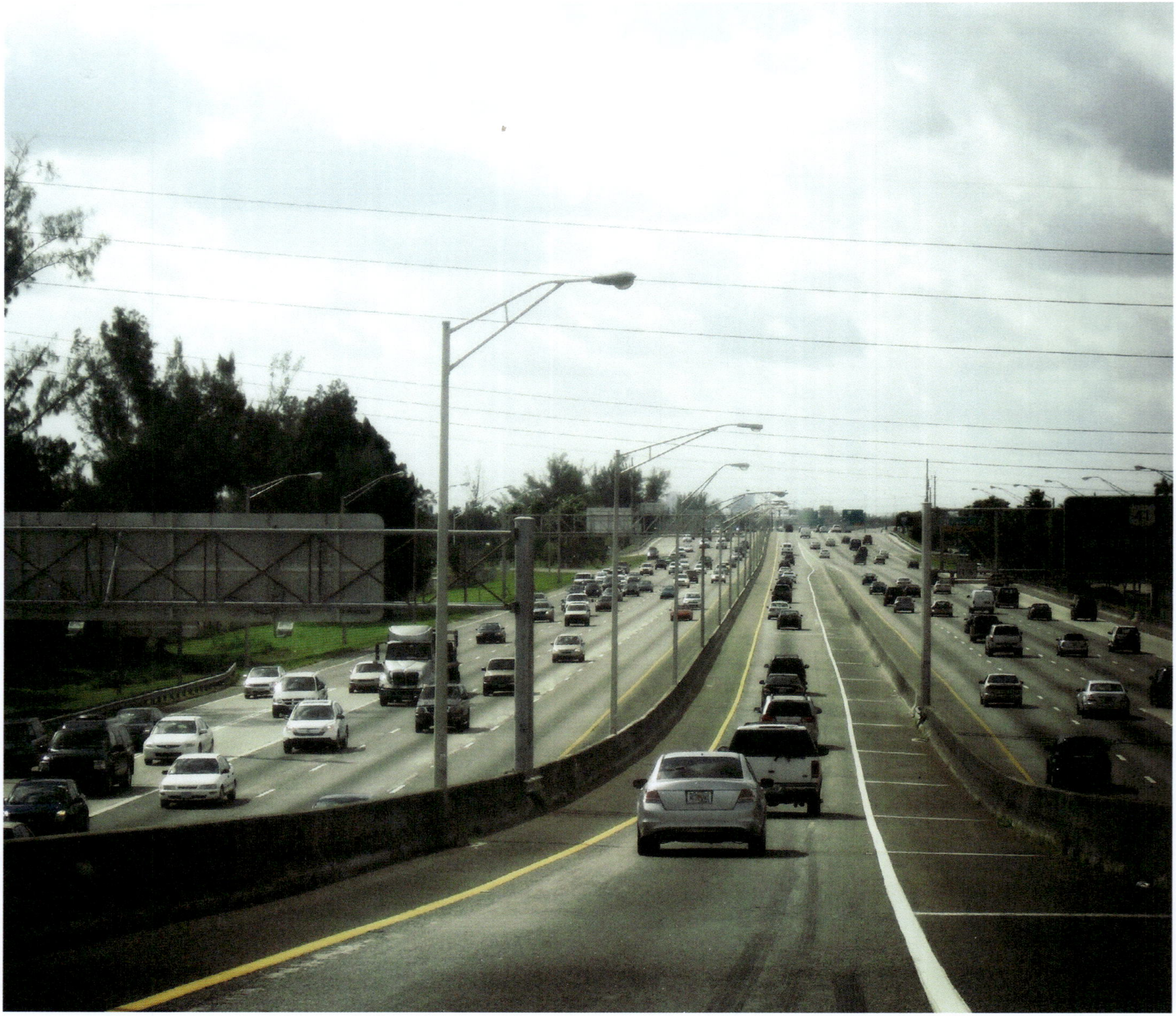

Life is not simple.

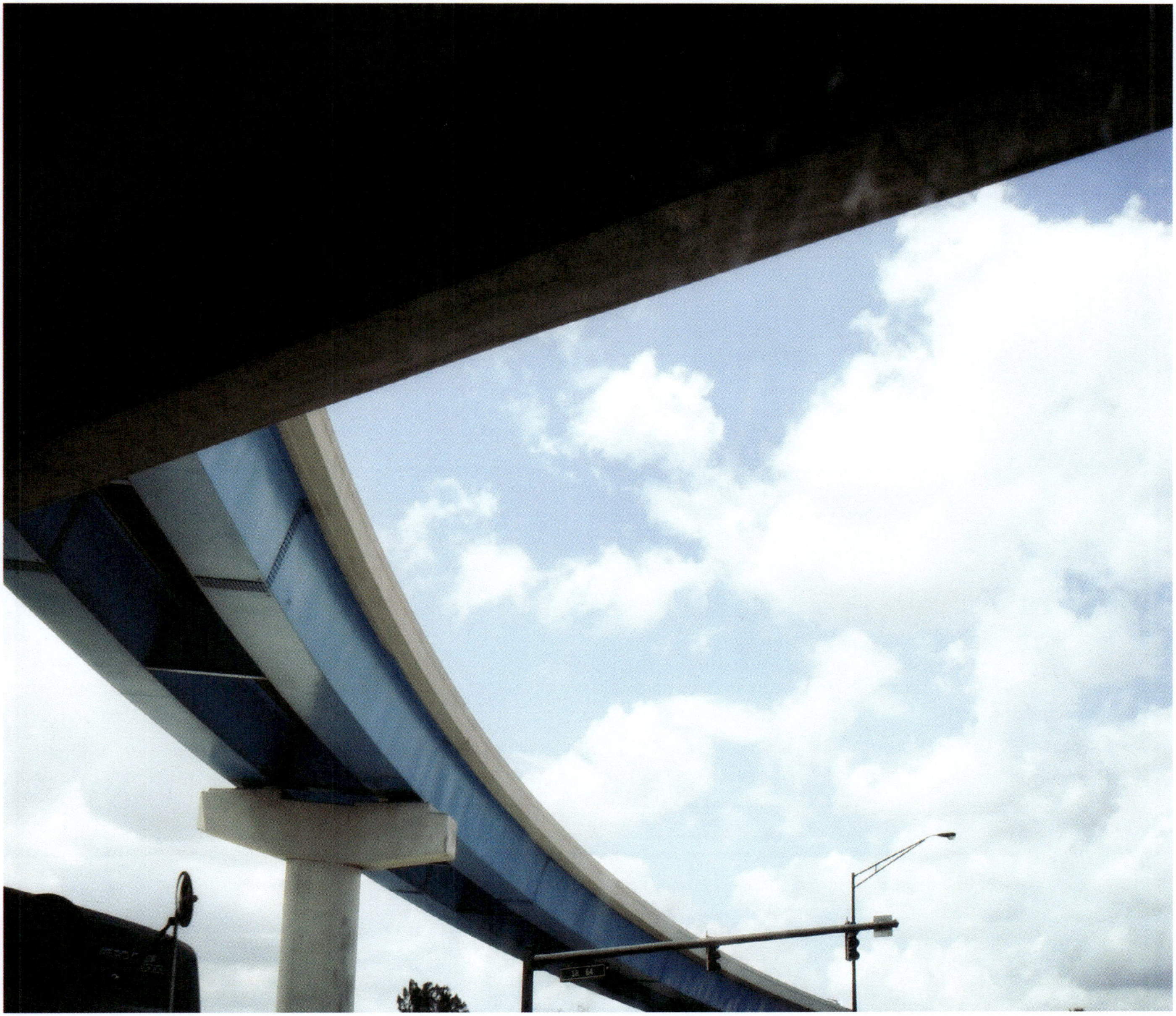

Blue skies are always over some bridges.

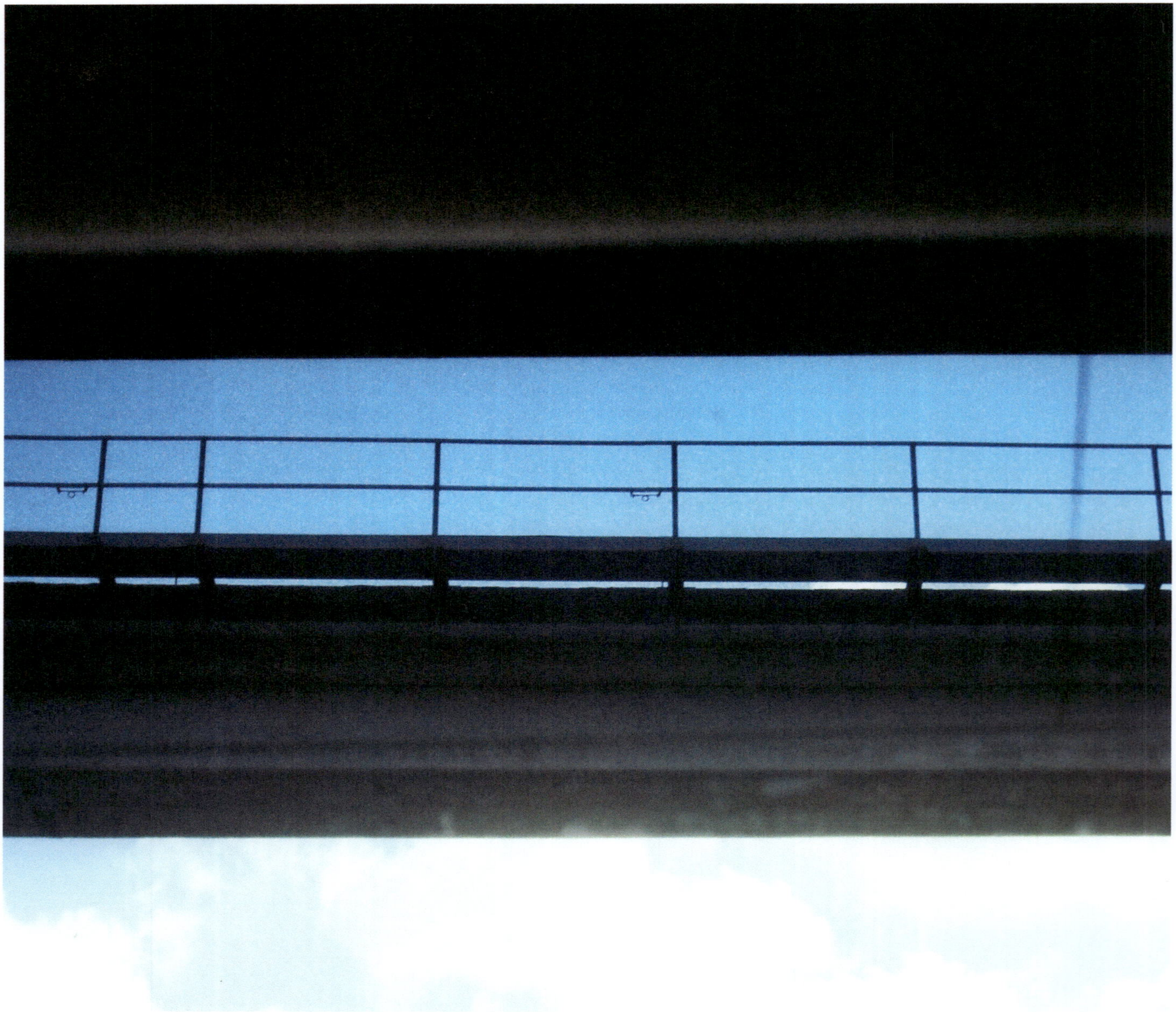

Some see beauty that others don't see.

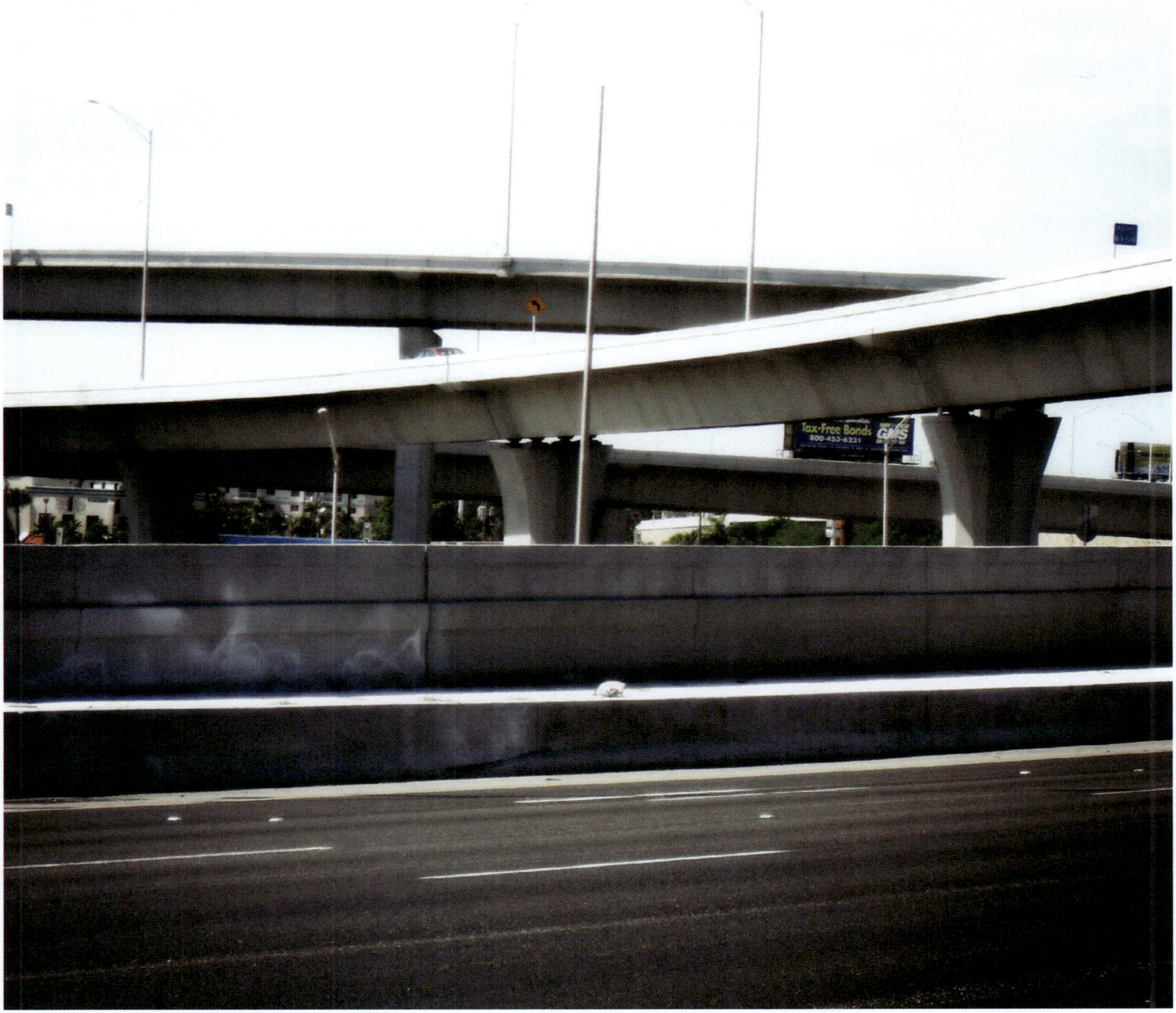

When opposites unite, harmony is created.

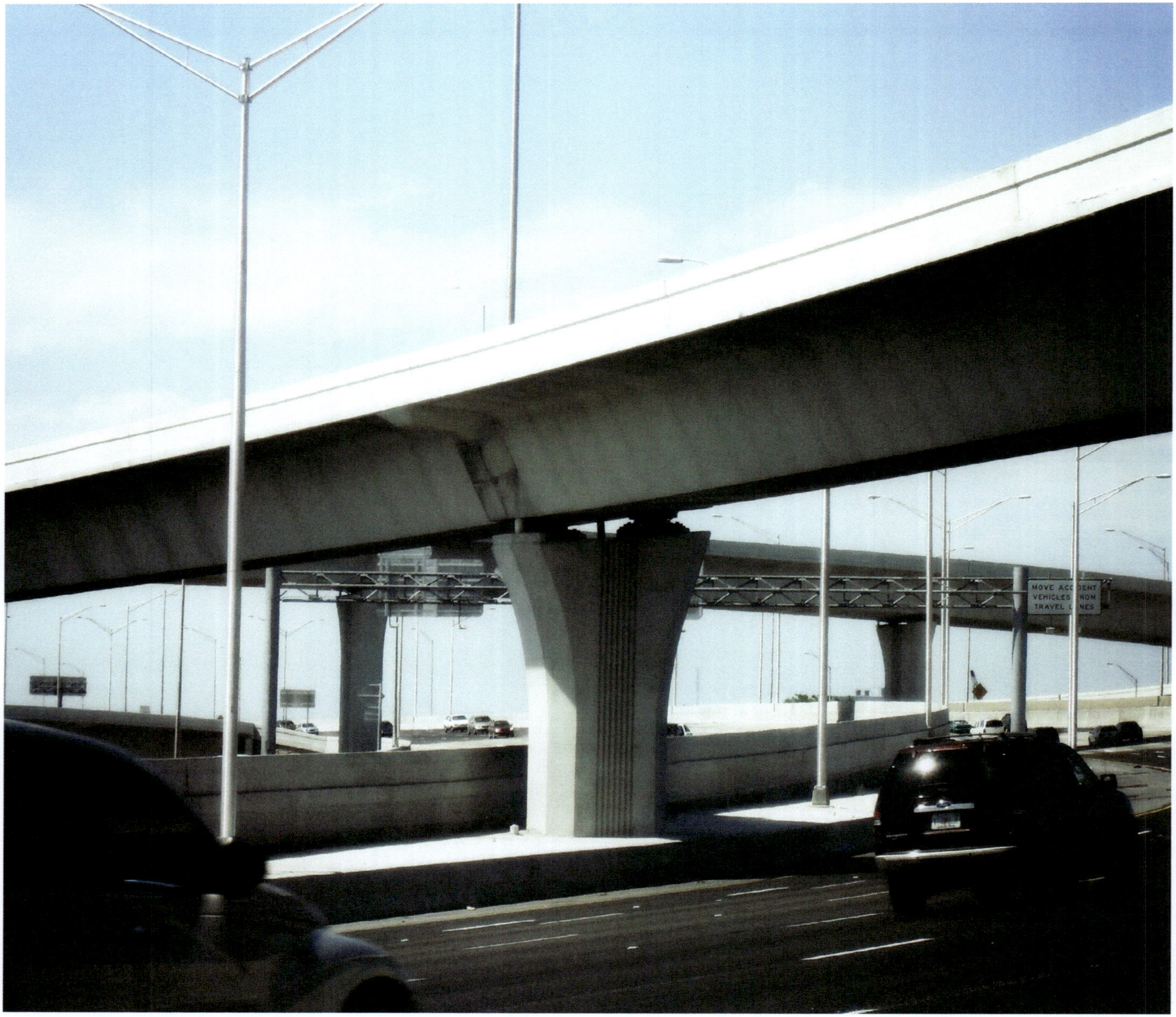

Engage in the process.

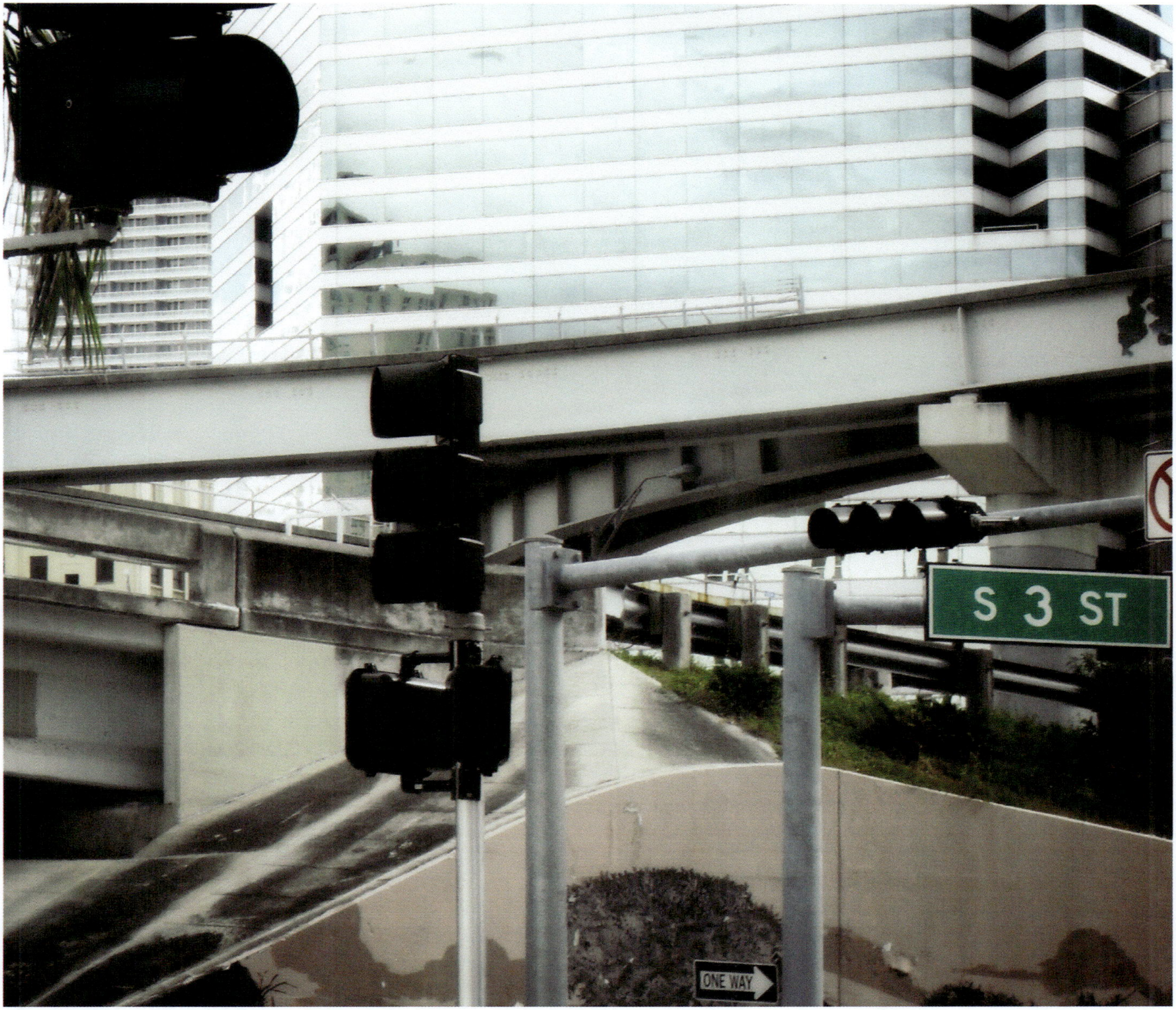

Being open and curious allows the world to unfold.

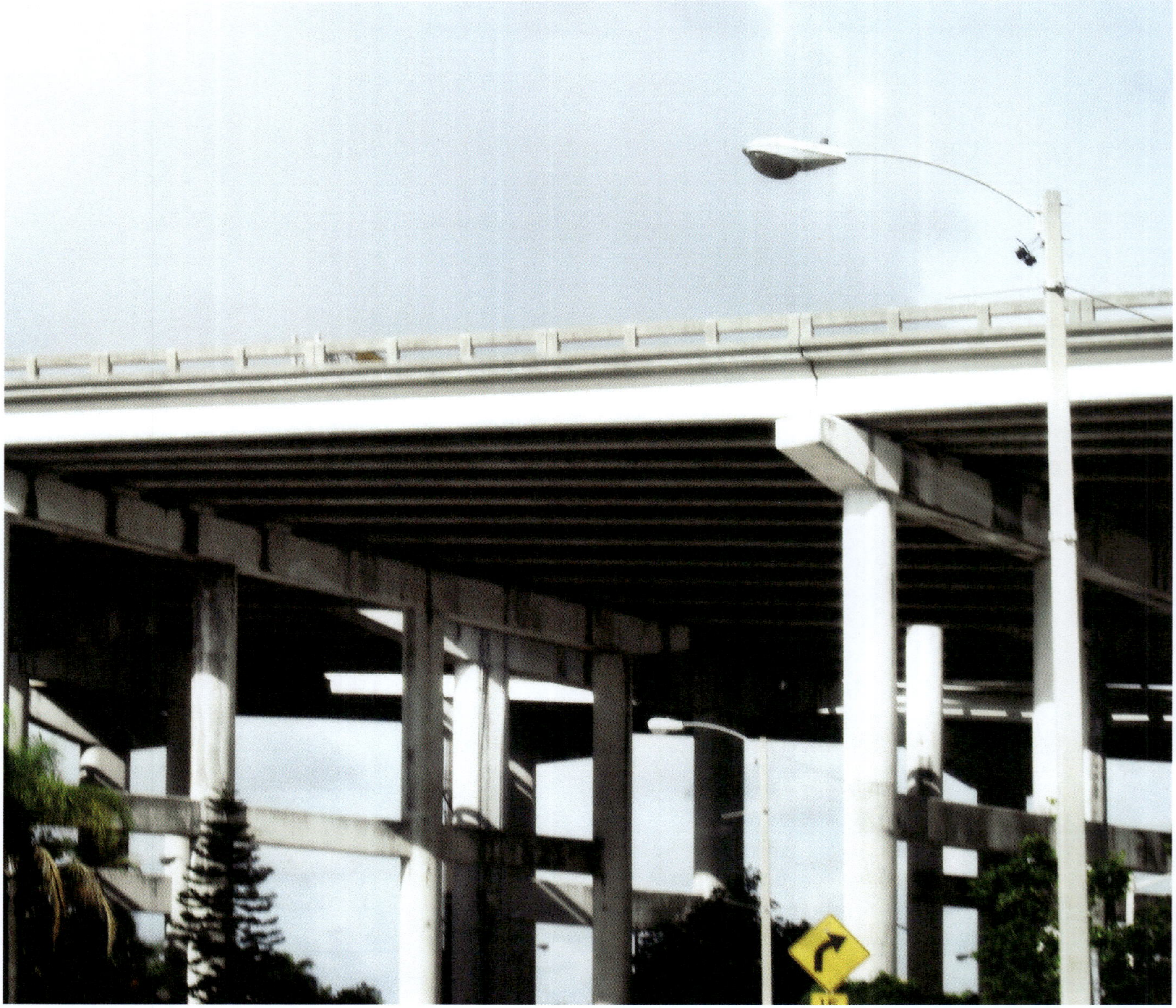

New cities await.

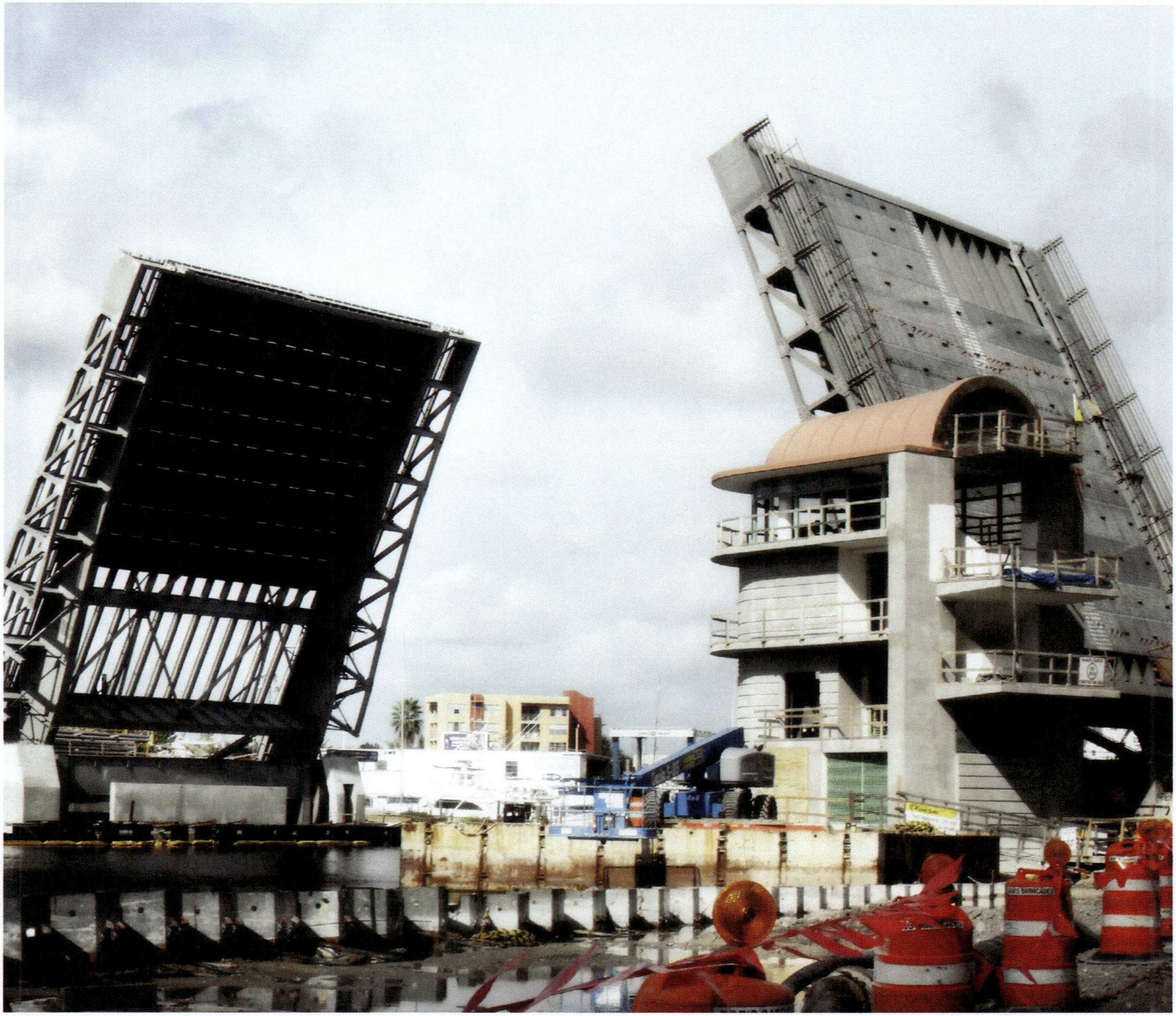

The bridge opens and allows passage.

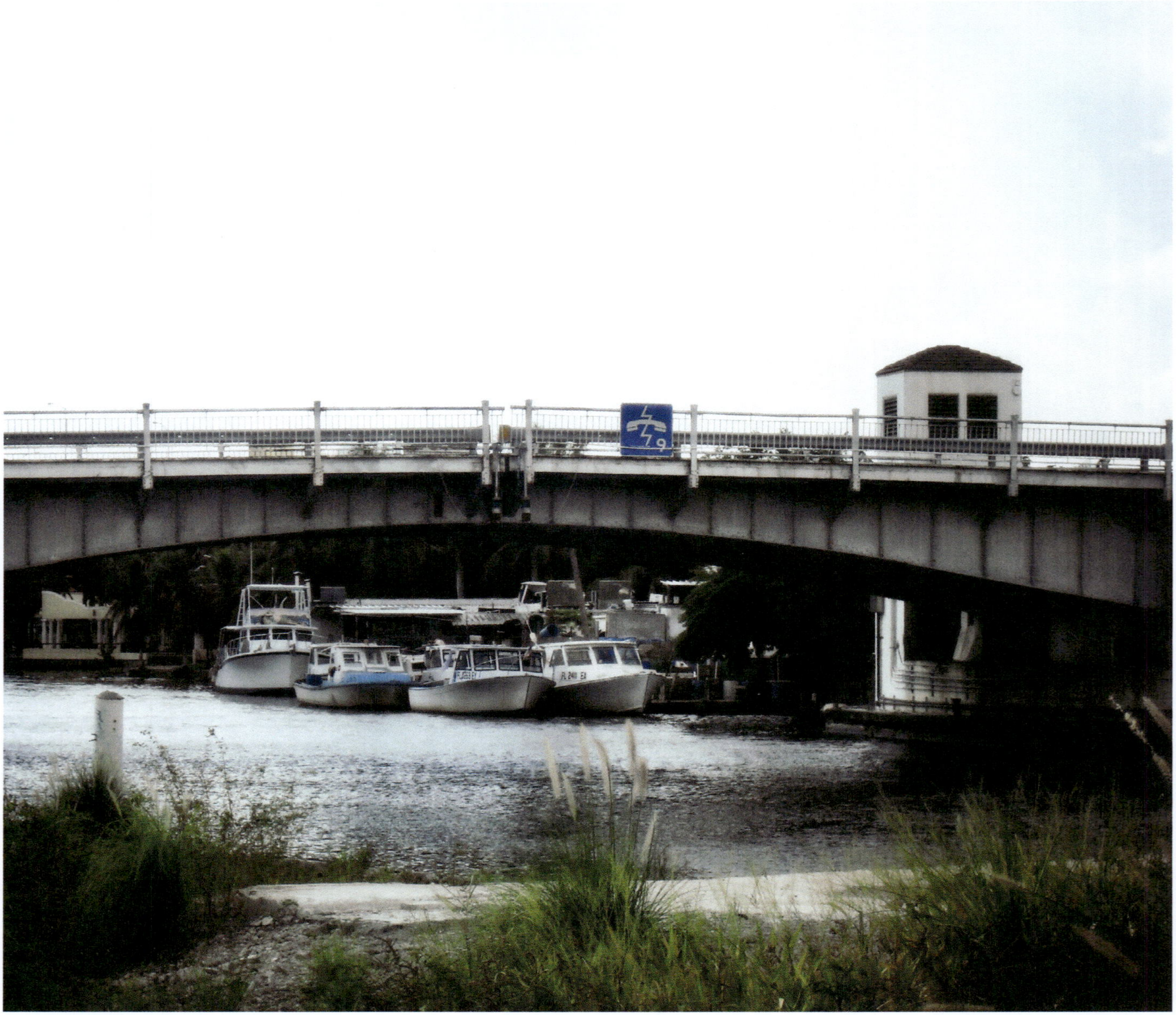

The bridge may be the only path to the other side.

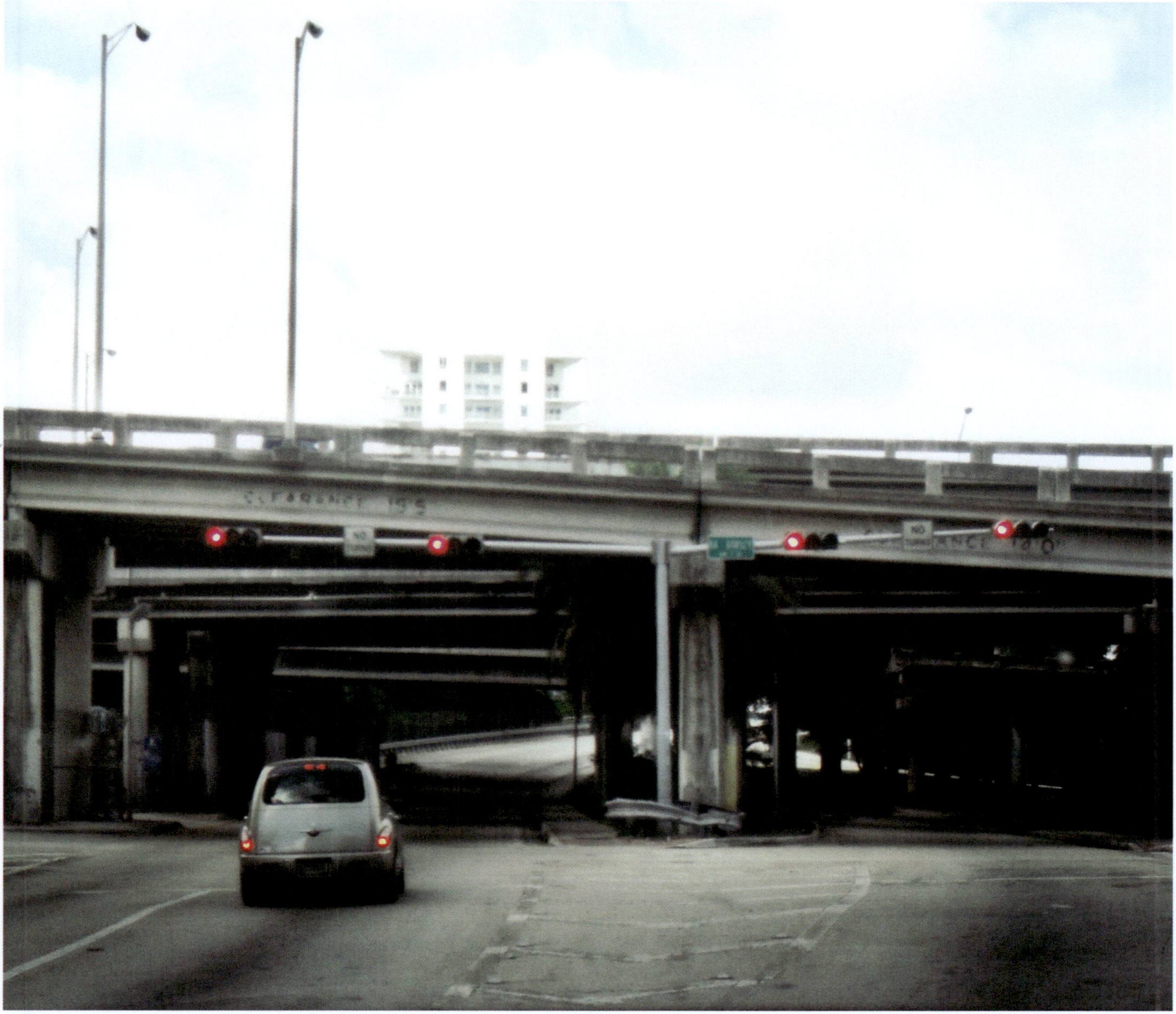

In conflict be fair and generous.

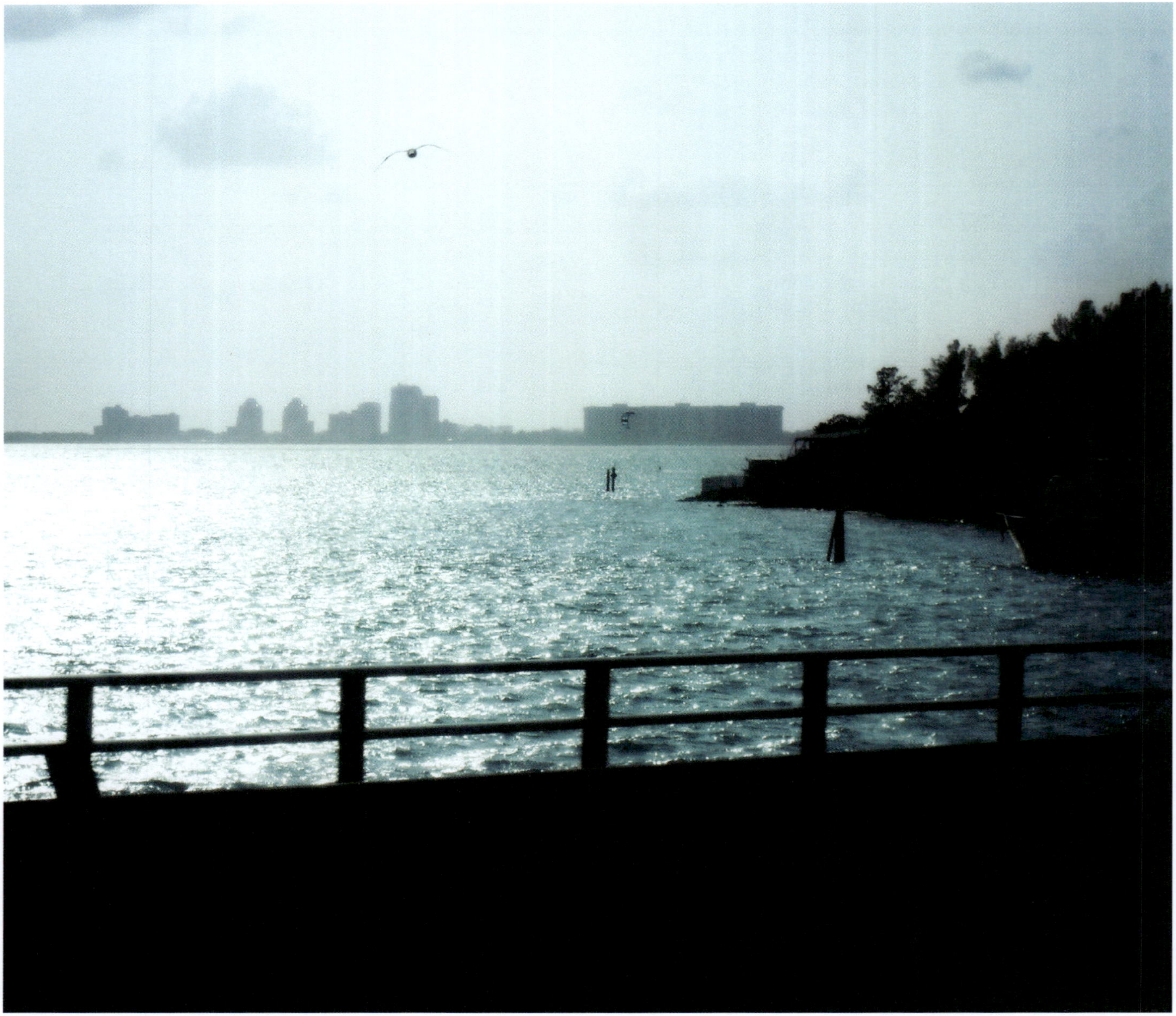

Cross the bridge to serenity.

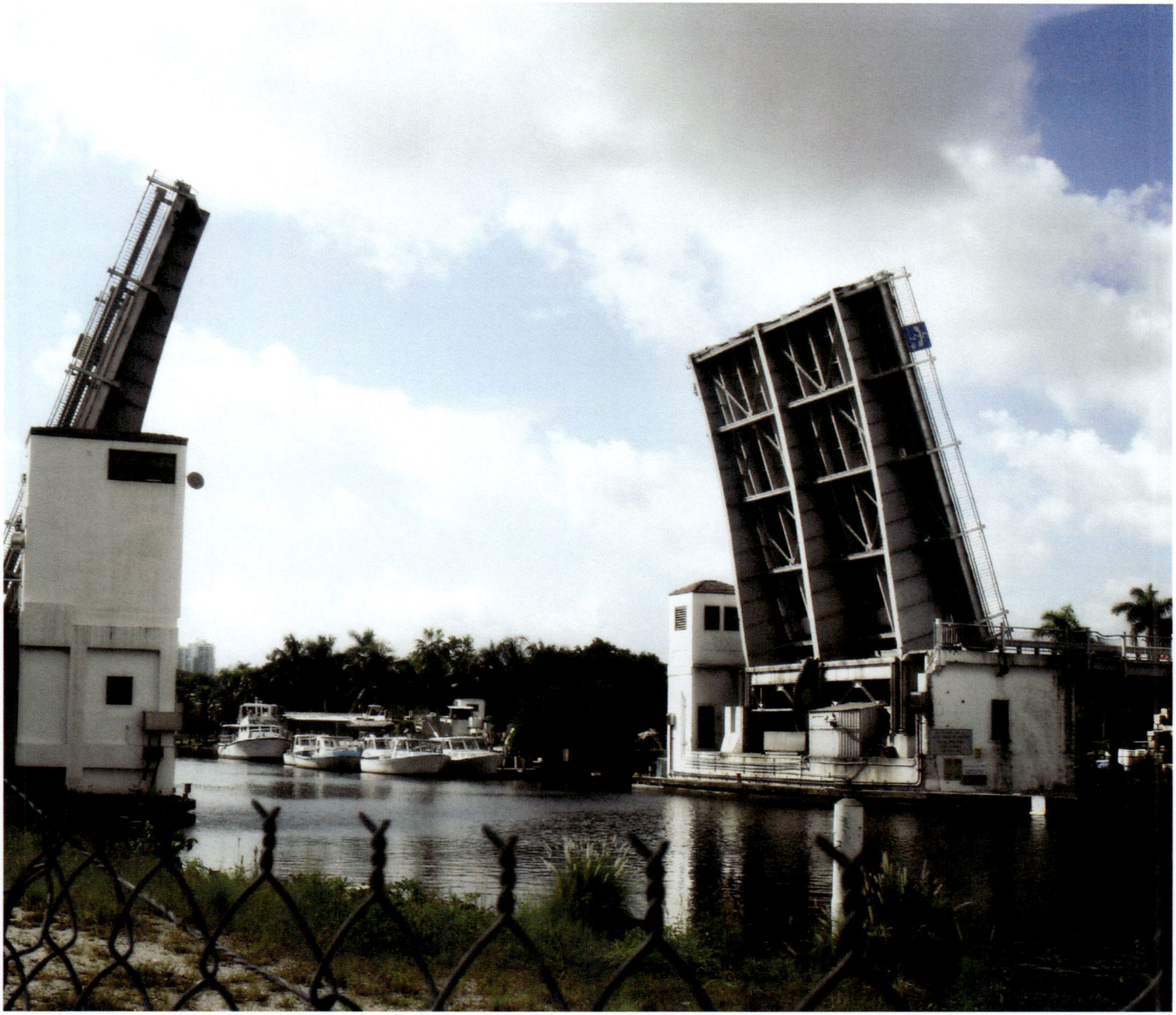

Bridges are keepers of the waterways.

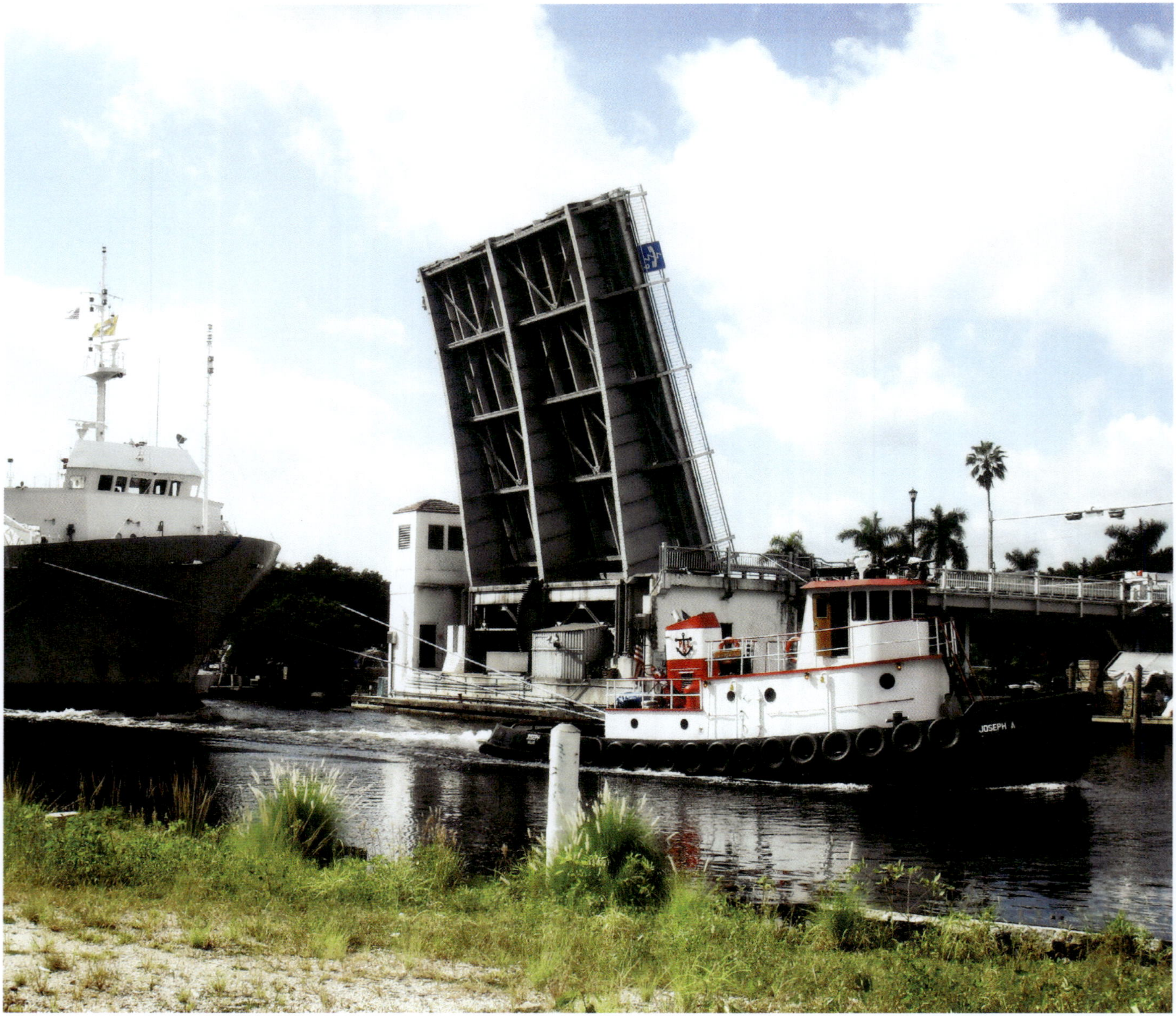

Stay in the center and all things will take their course.

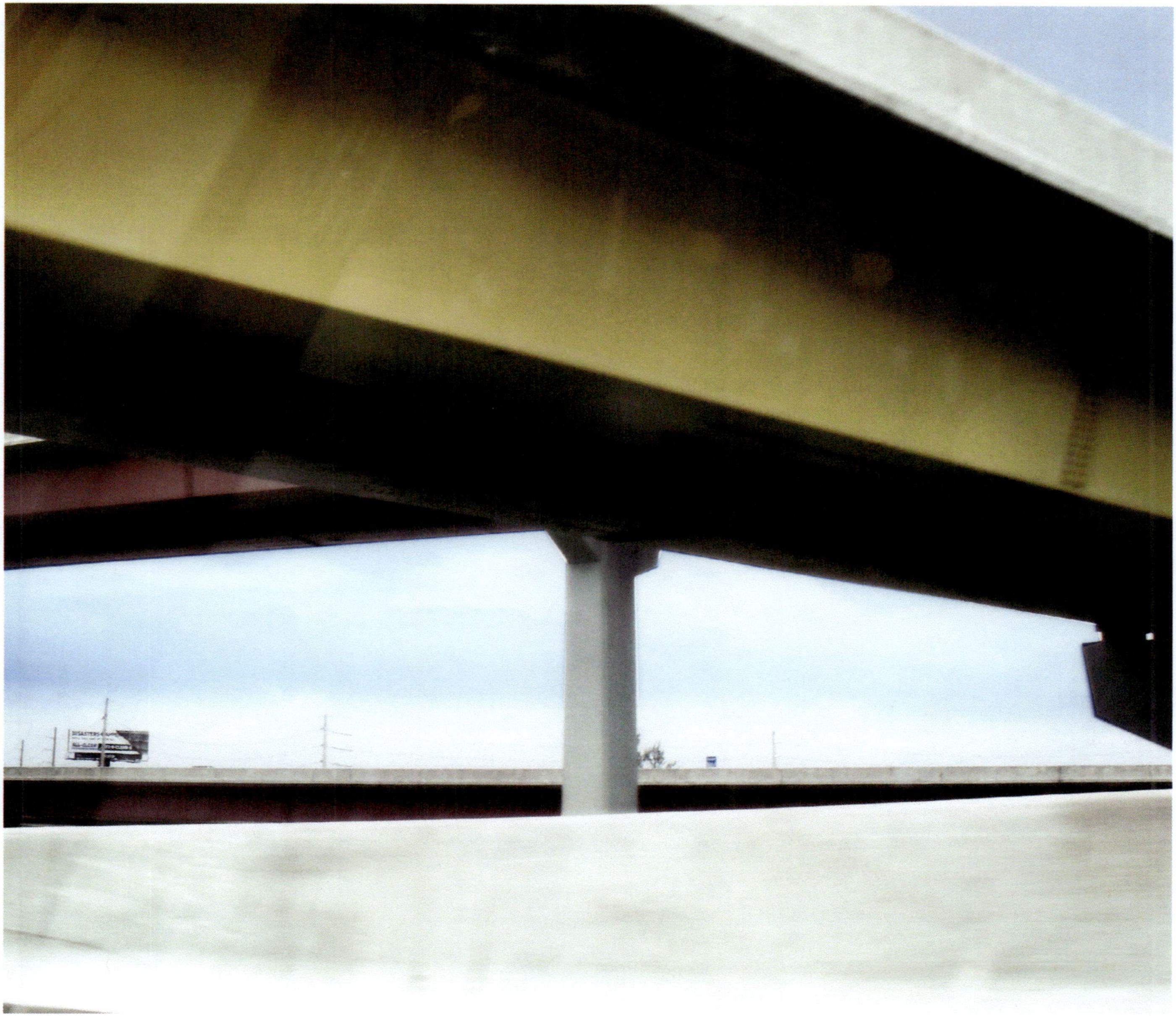

For many it's the journey, the transition, not the end that is important.

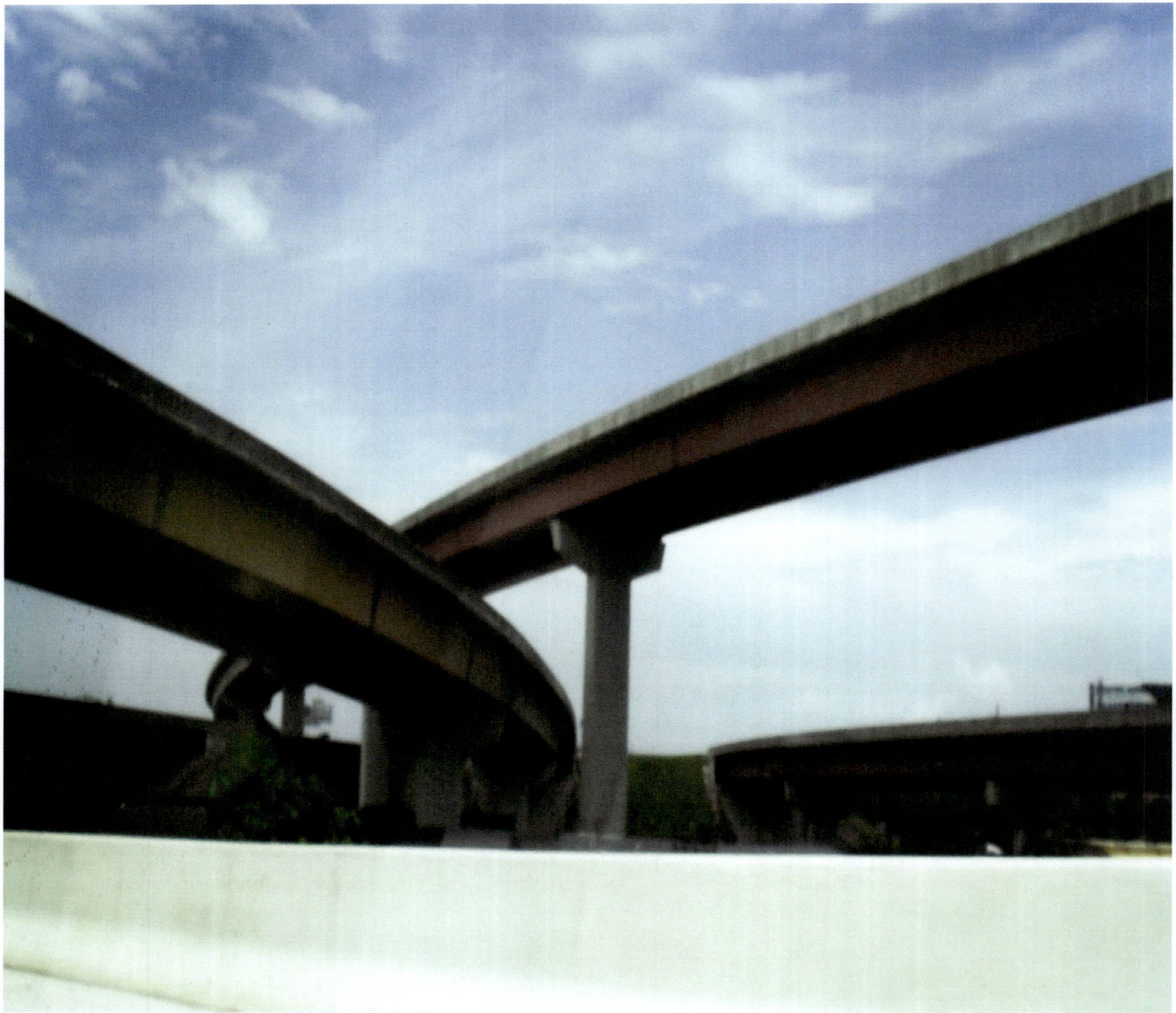

Partners in different directions.

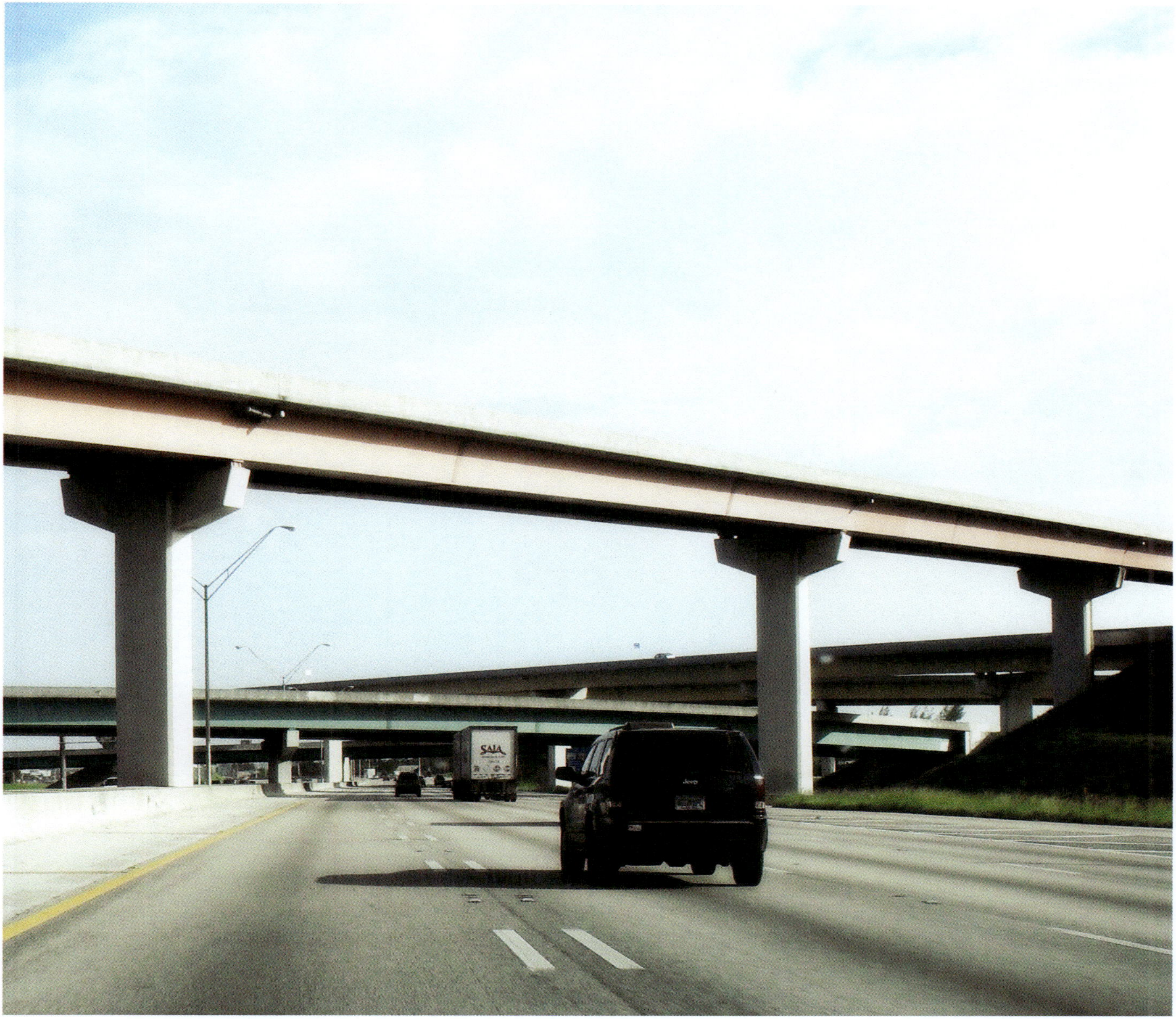

Is it the end of the bridge or the beginning?

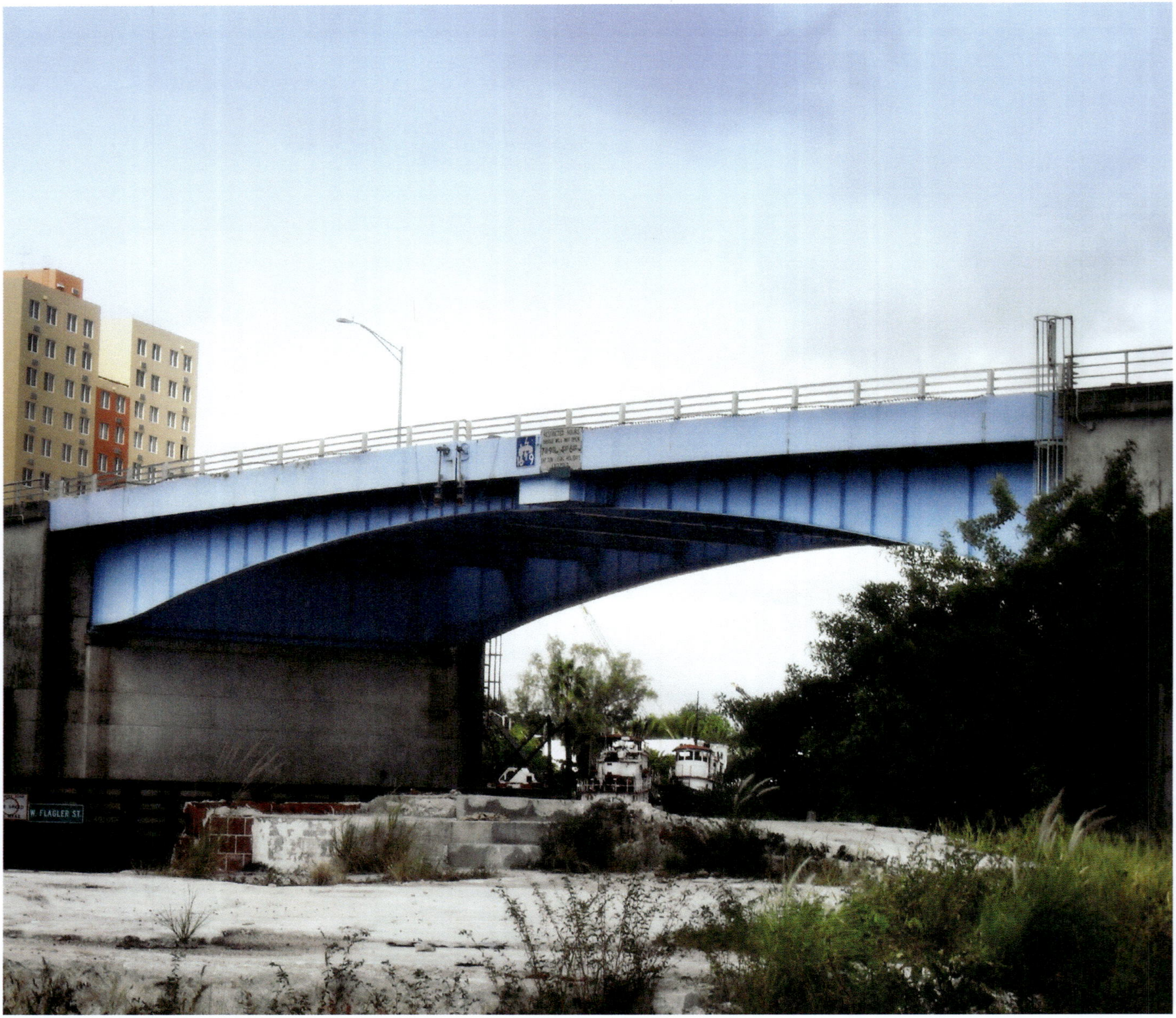

There is a time to be closed.

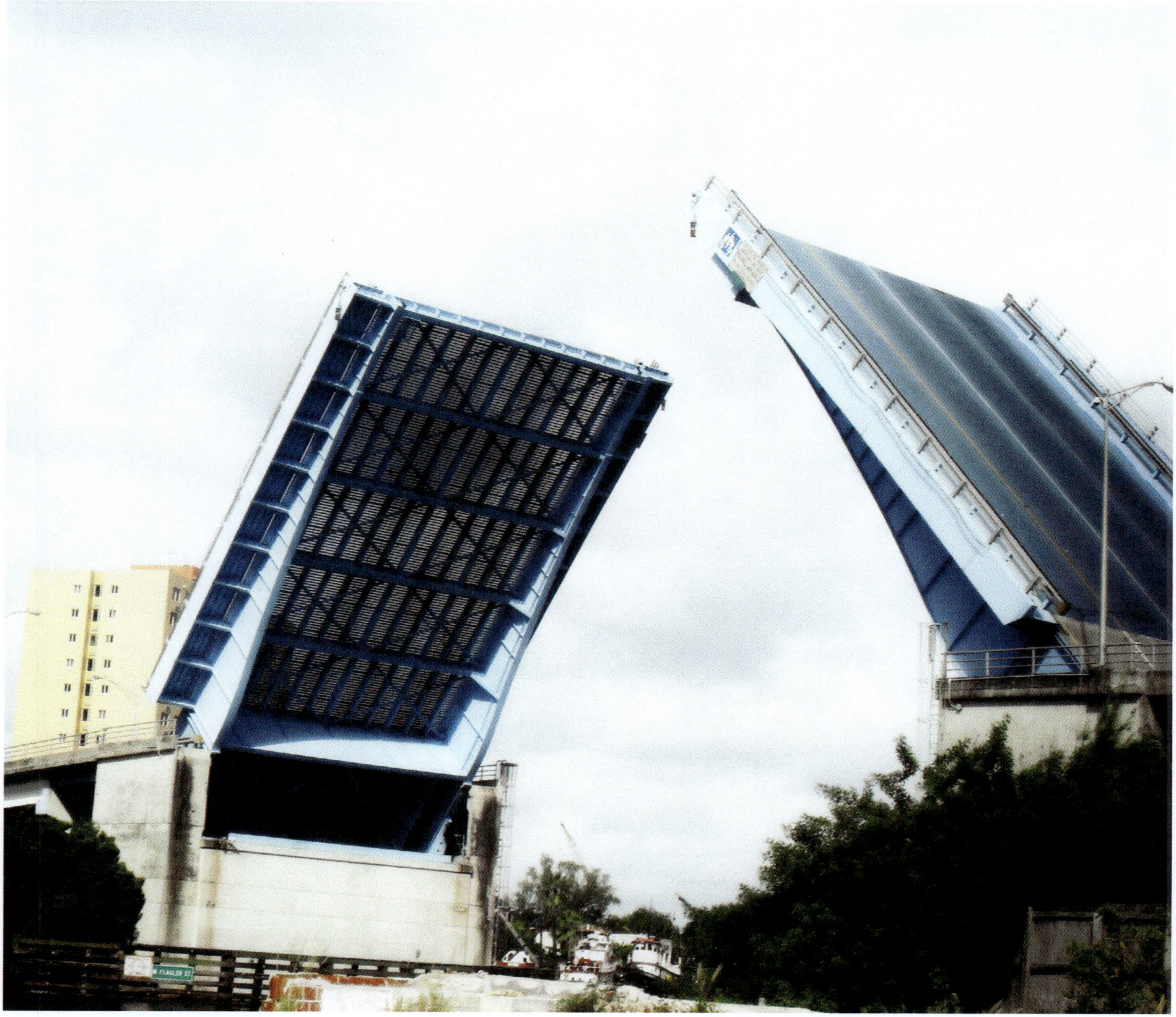

There is also a time to open up.

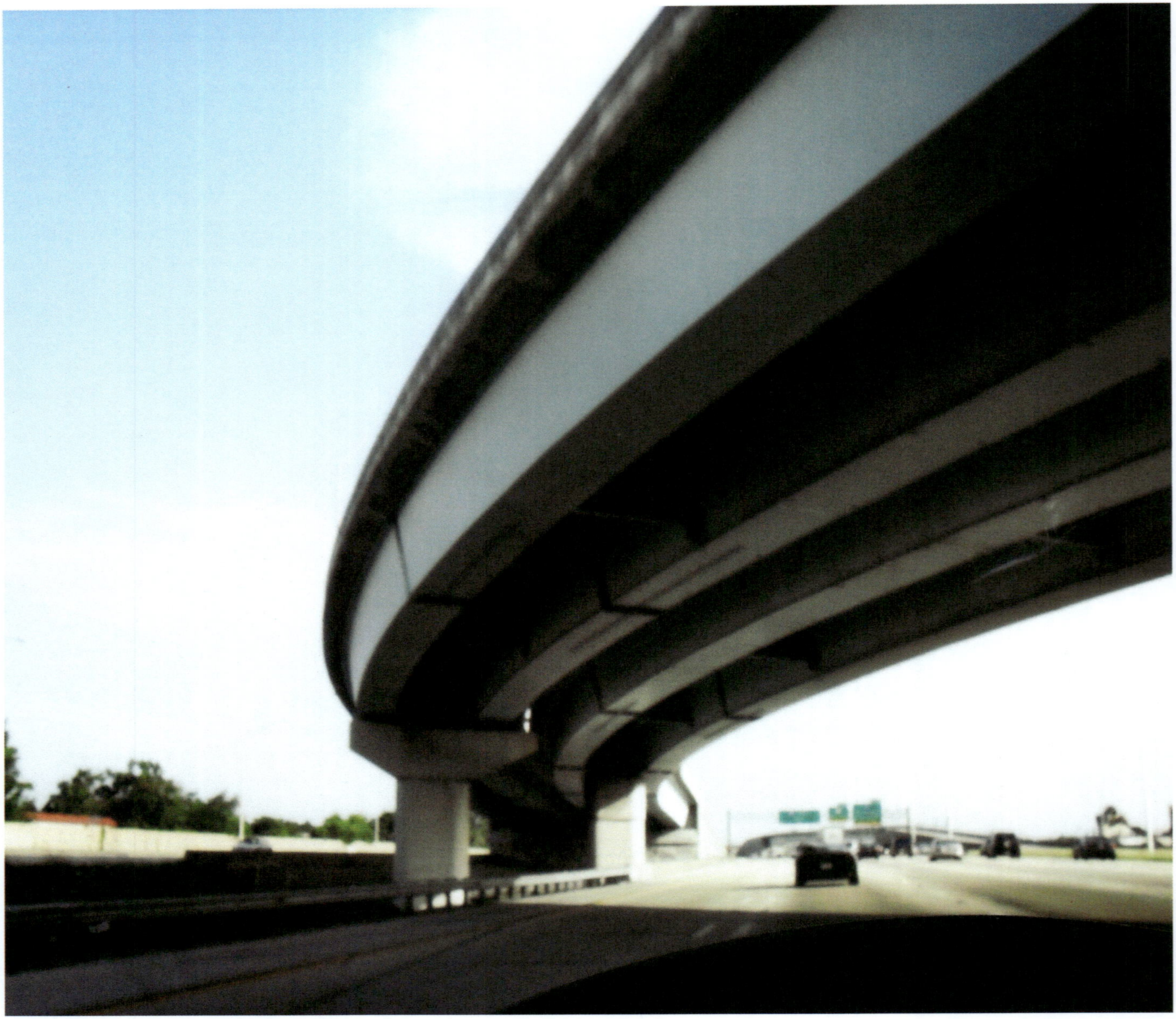

A place to speed after money or fame.

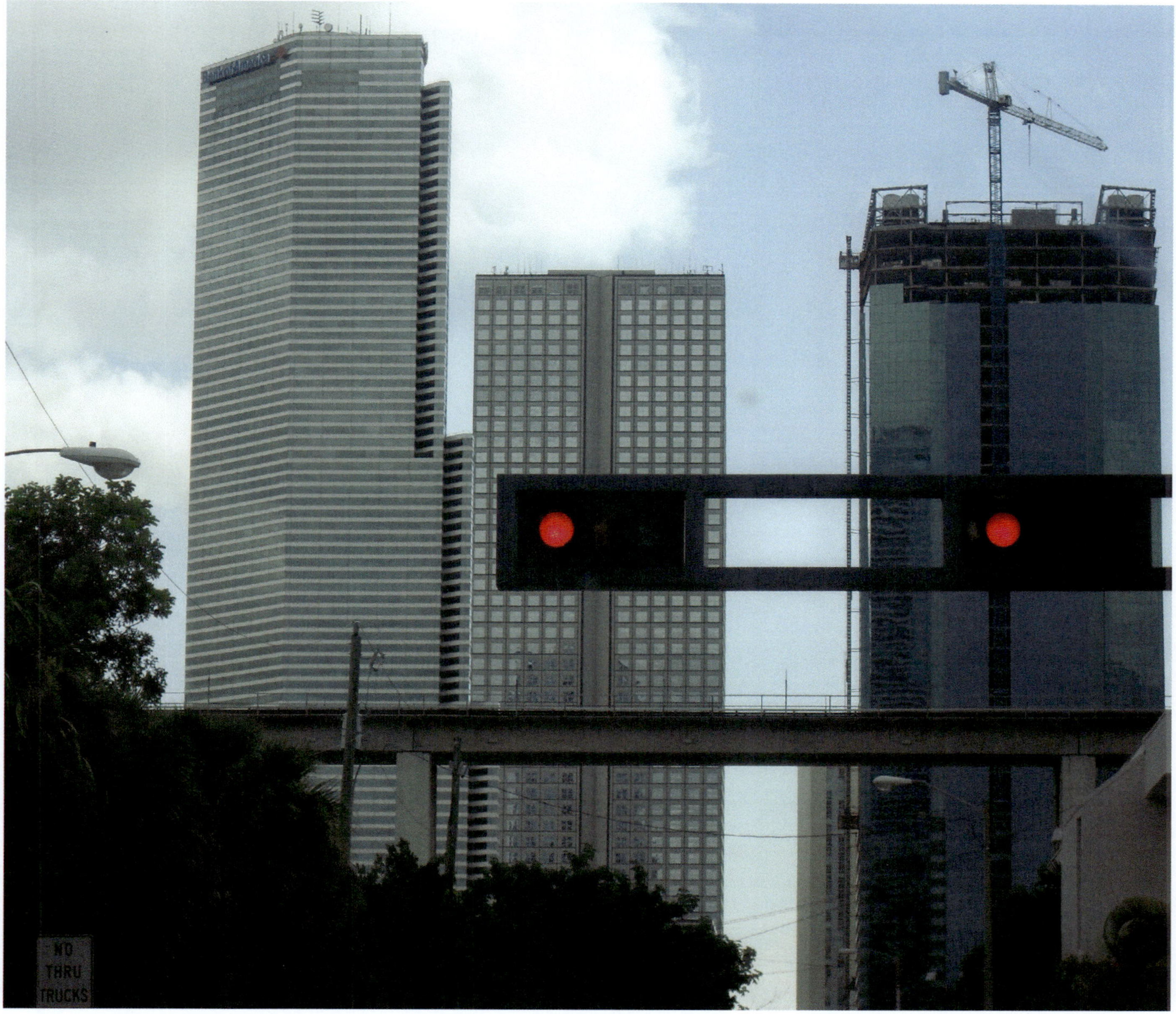

There is only change.

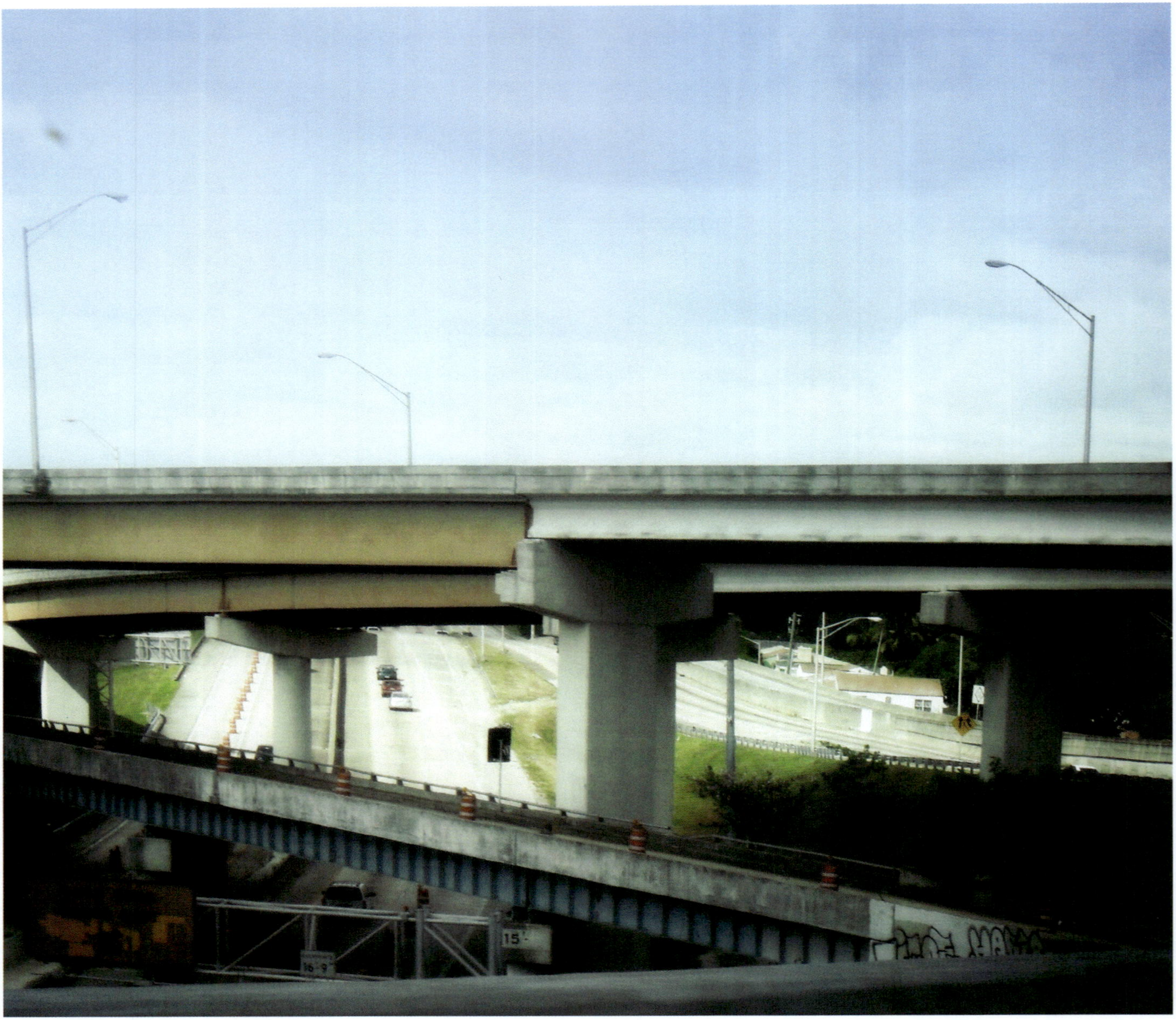

The pillar is the rock of the foundation, the parents of the structure.

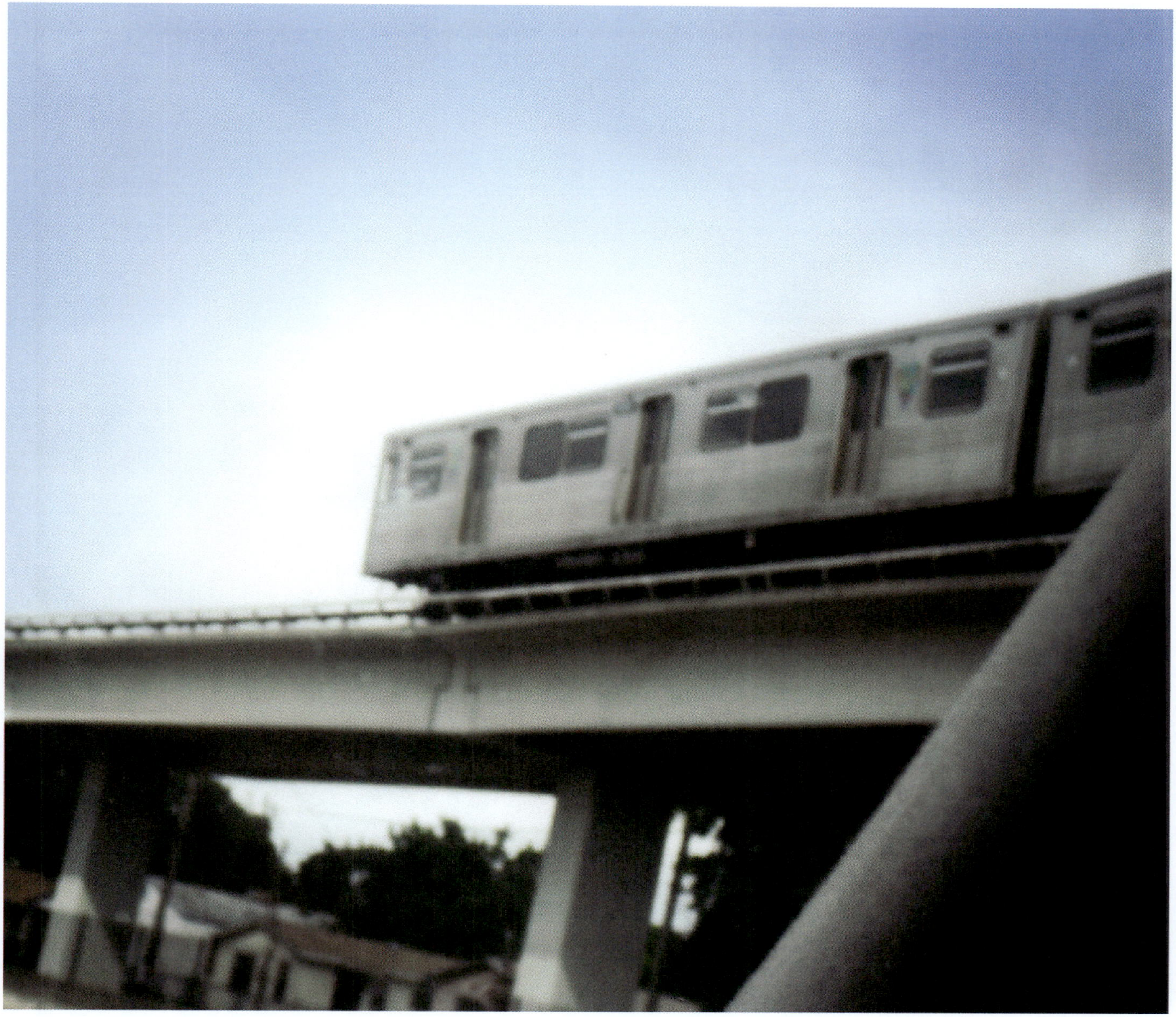

Trains, trucks, buses and cars are served by bridges.

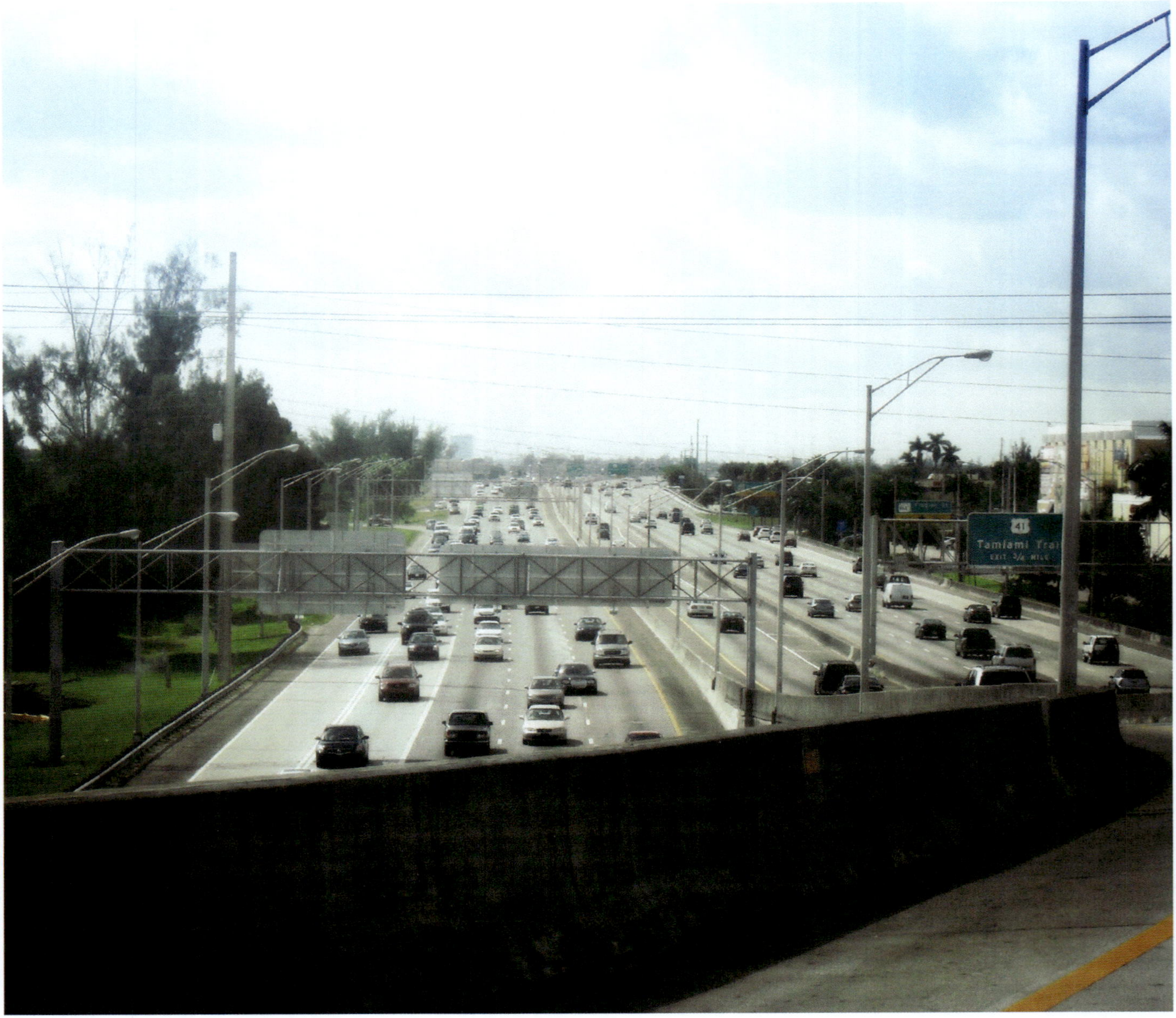

If you close your mind in traffic with desires you will be troubled.

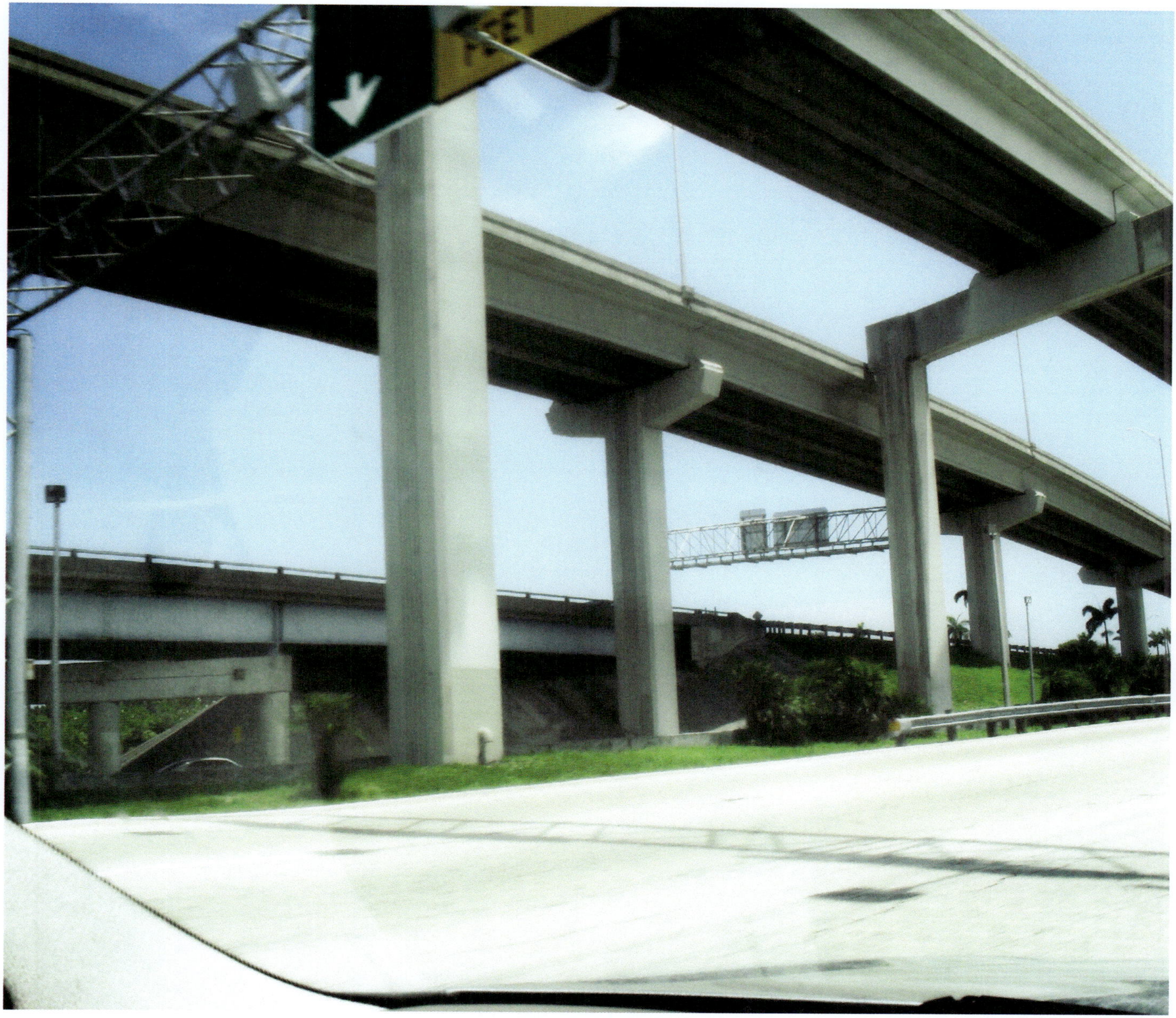

On your journey travel without effort.

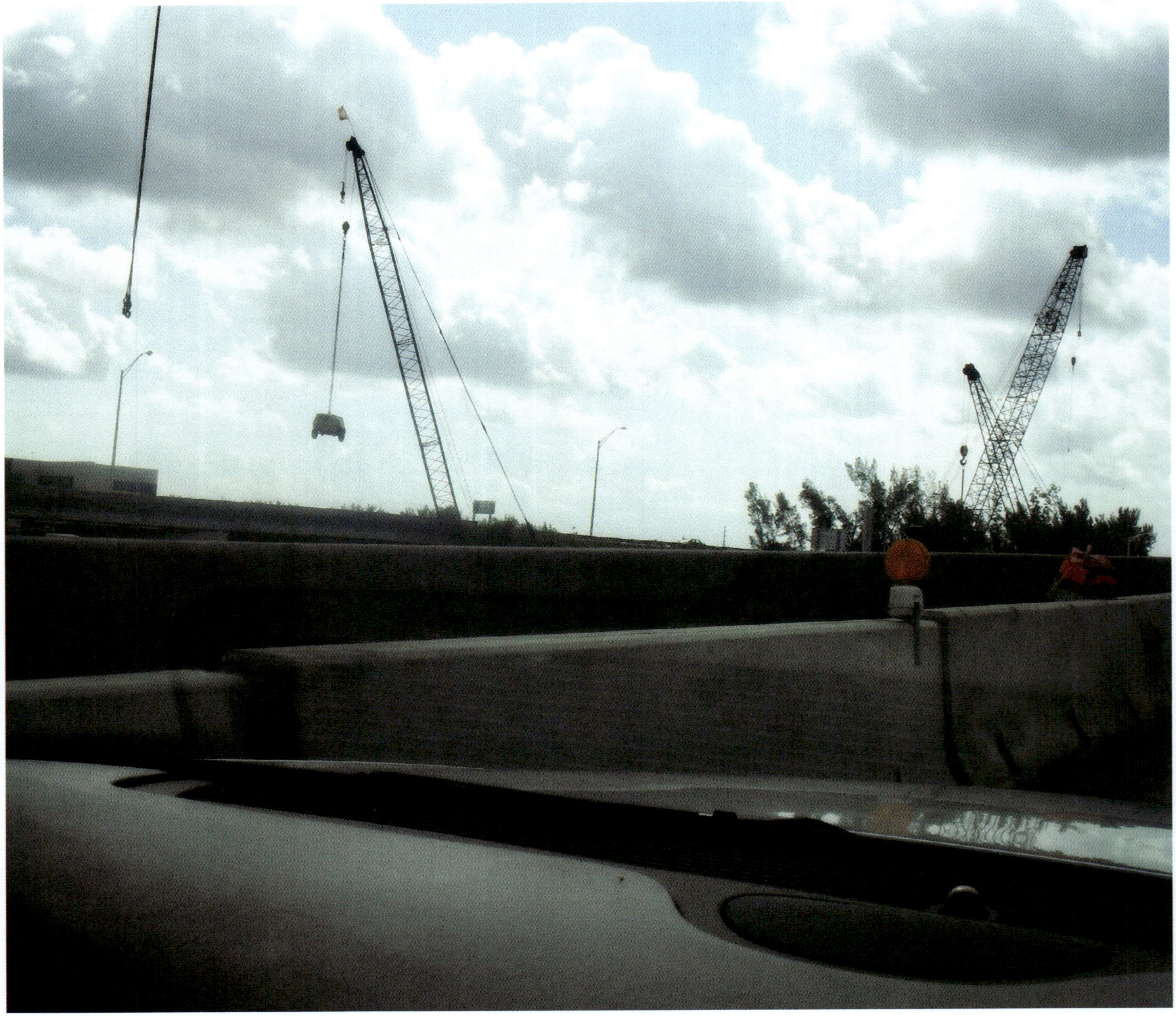

The system is constantly built and rebuilt.

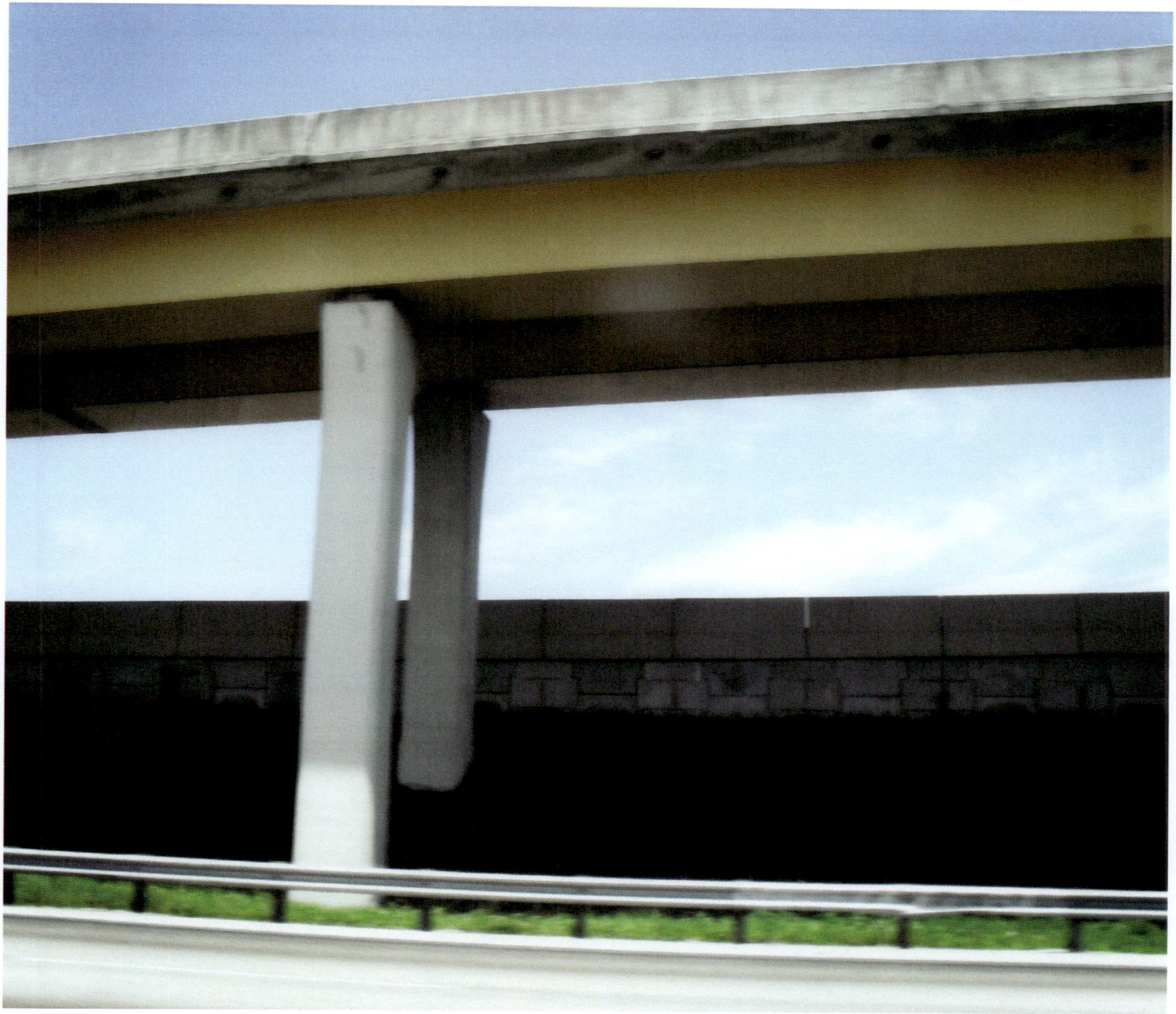

Take time to enjoy the beauty of the journey.

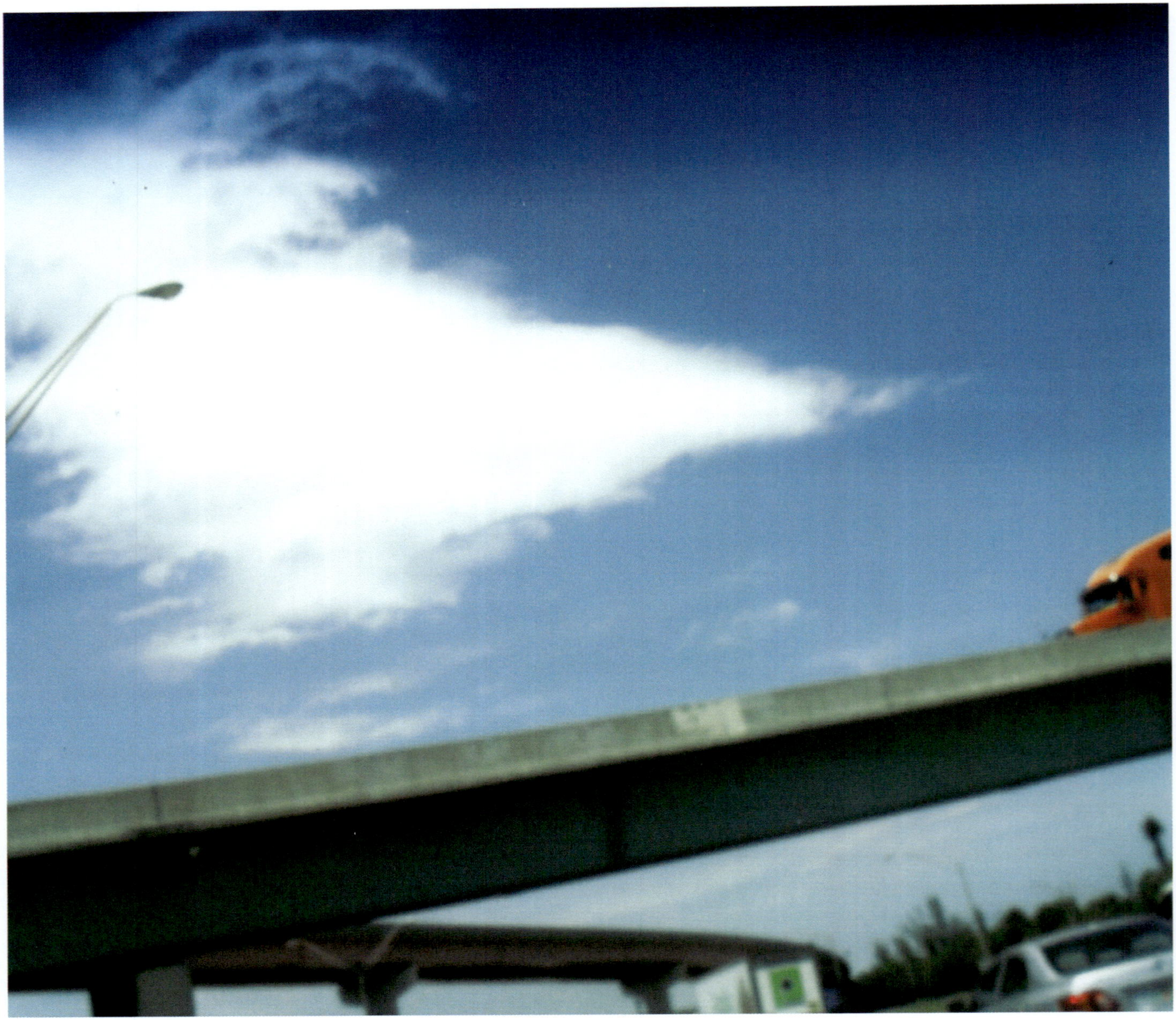

As we move the universe opens up.

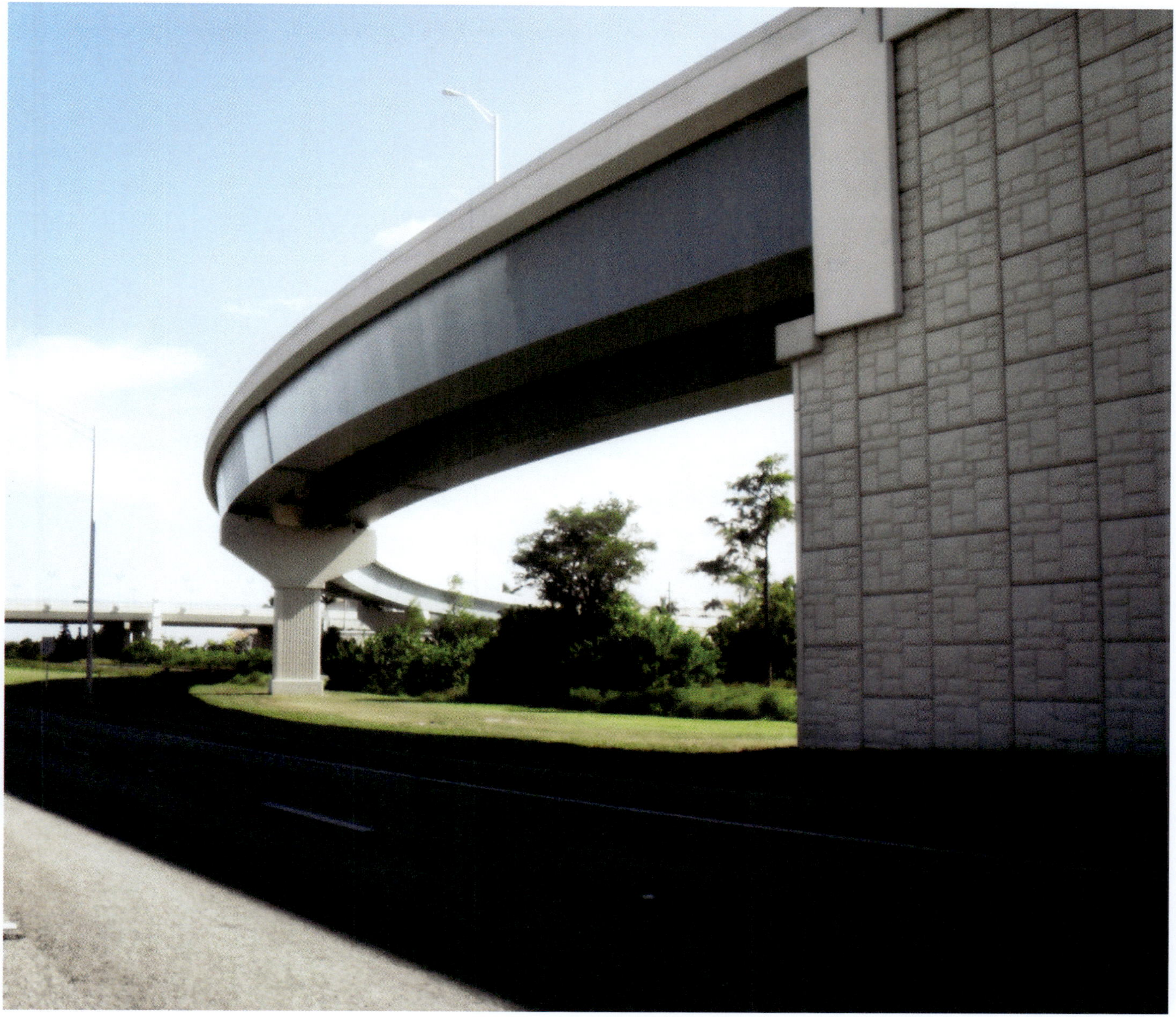

A time for being in motion.

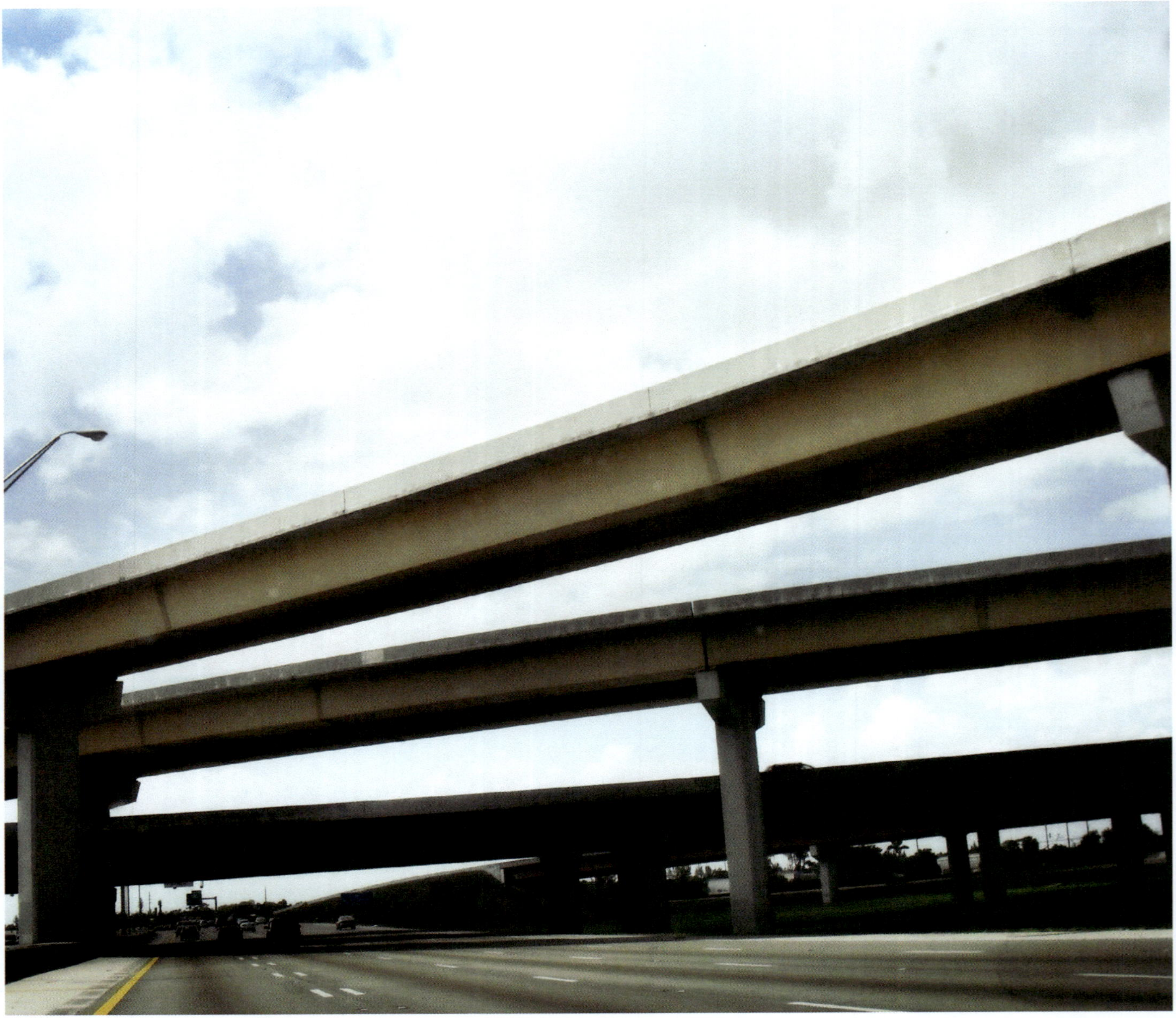

See the contrasts in all that you do.

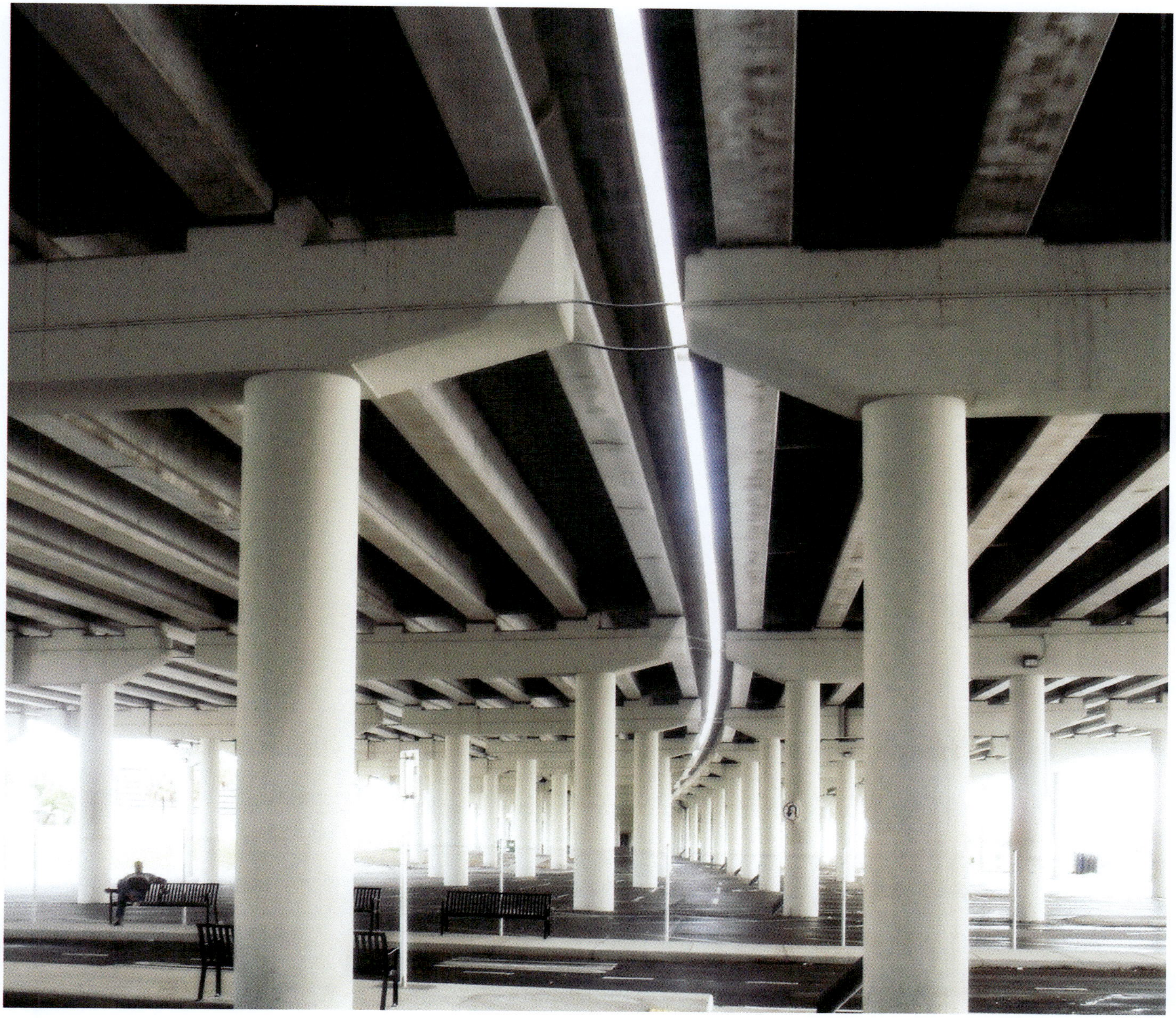

Everything is part of the other.

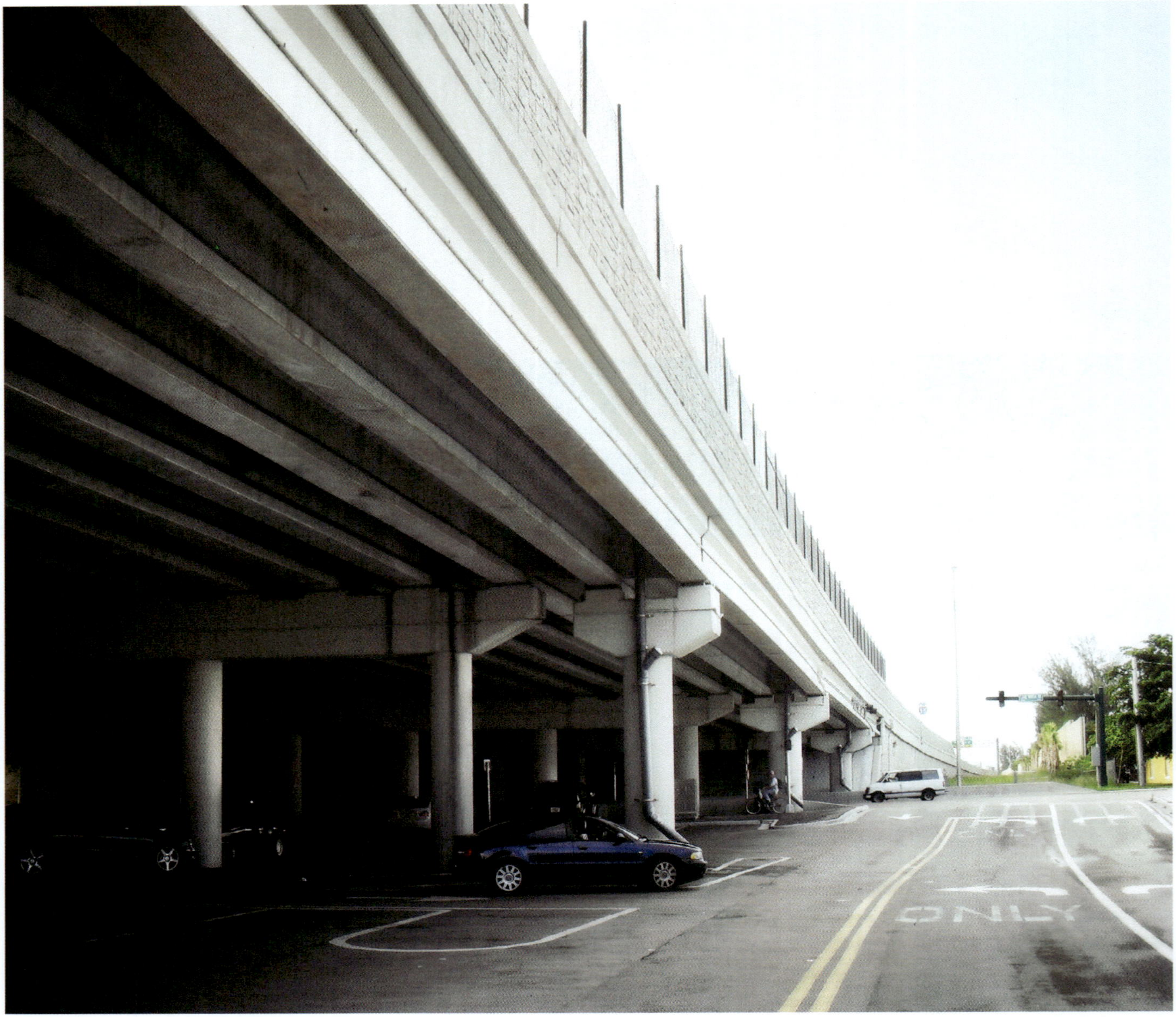

The road welcomes saints and sinners to its network.

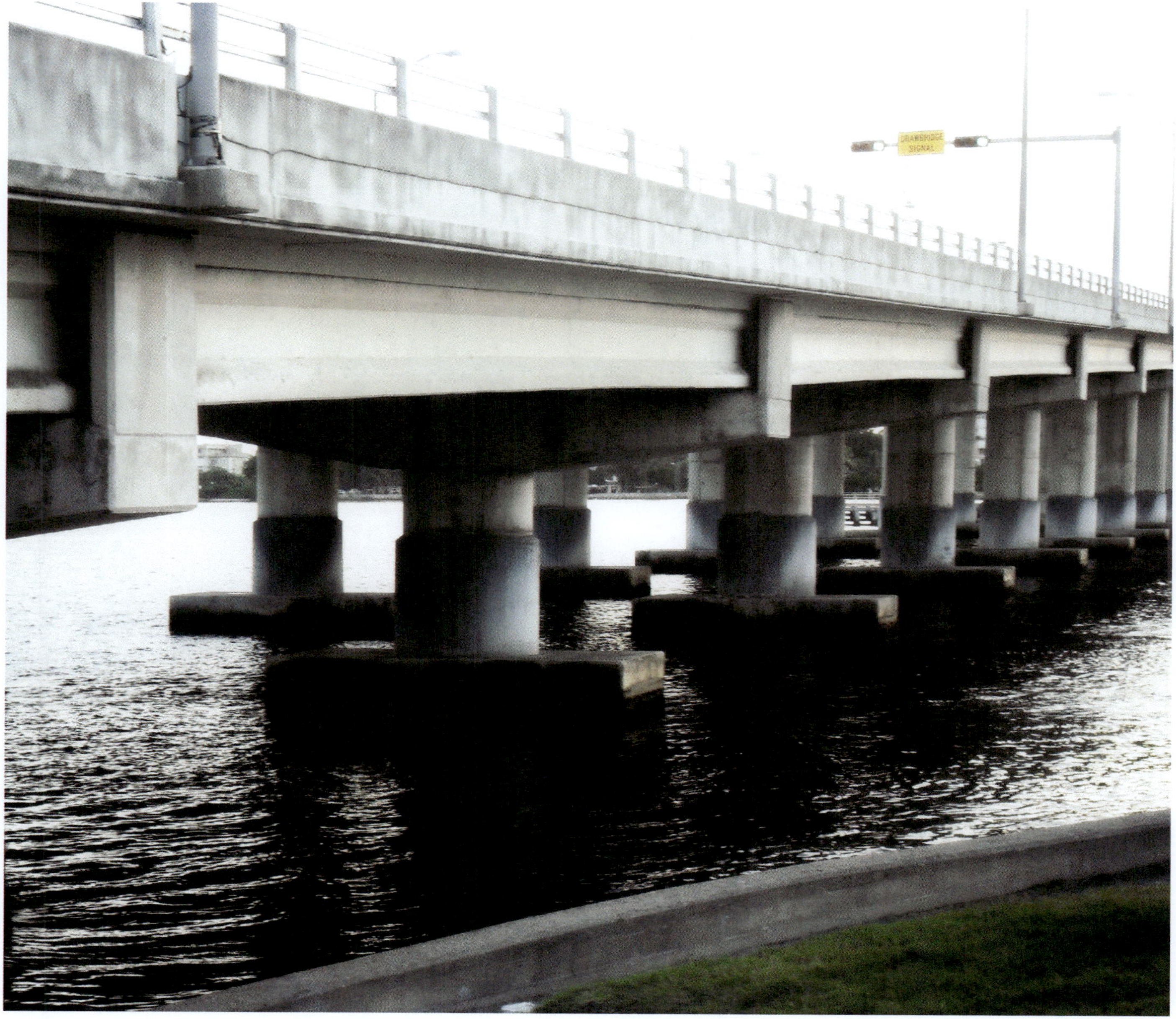

Cross a bridge to find a new life.

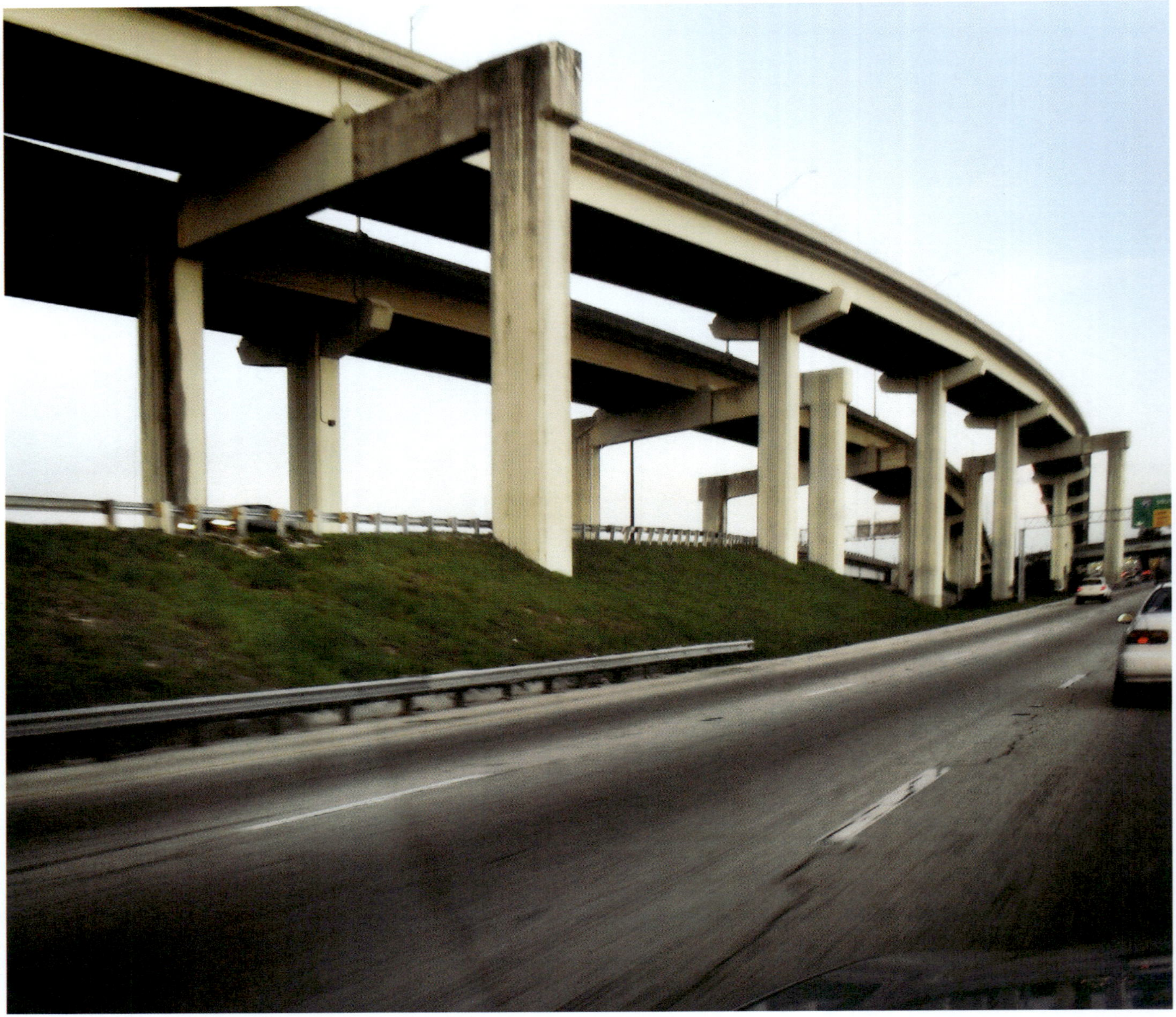

On the road one is always a stranger.

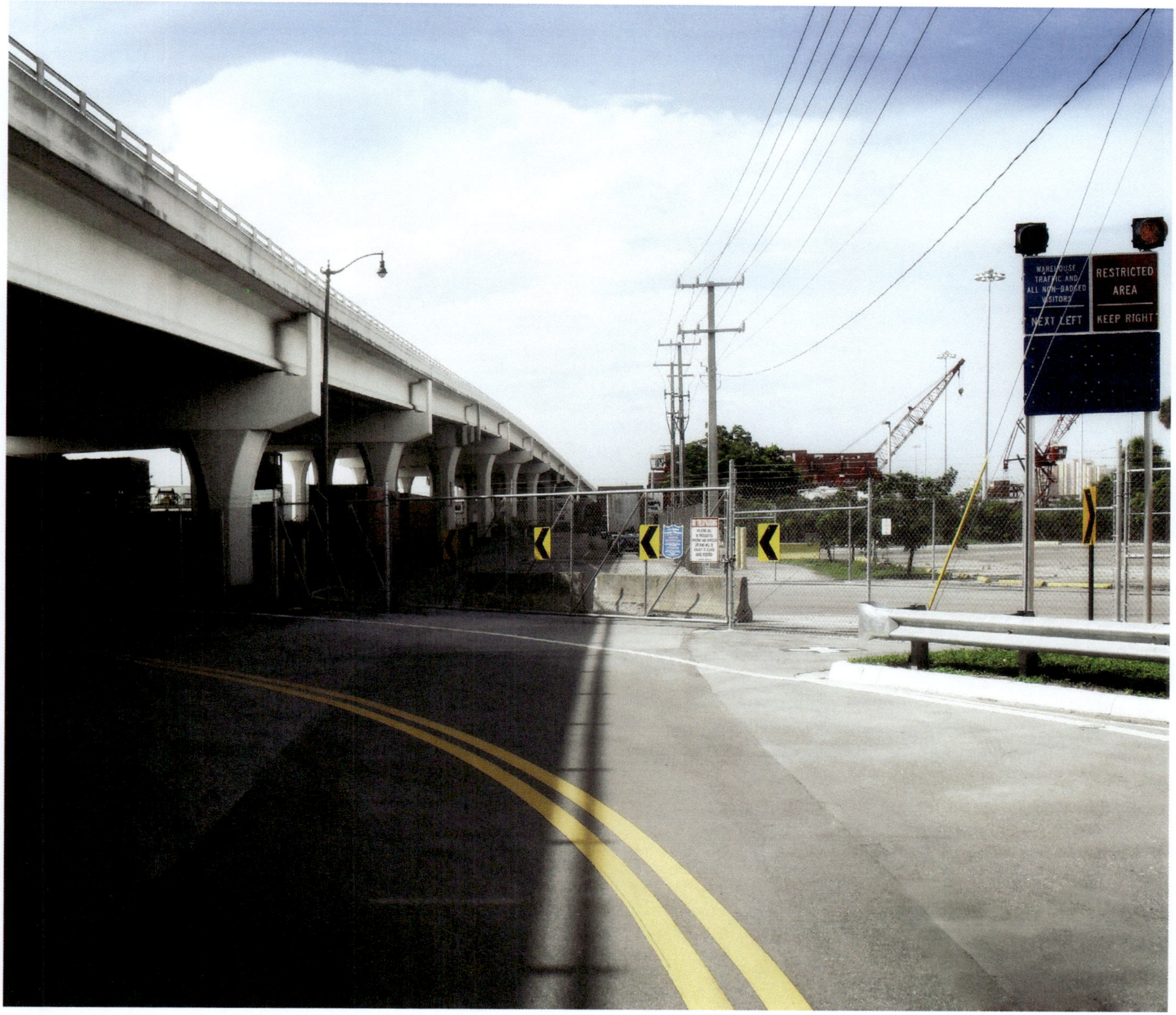

When nothing is done nothing is left undone.

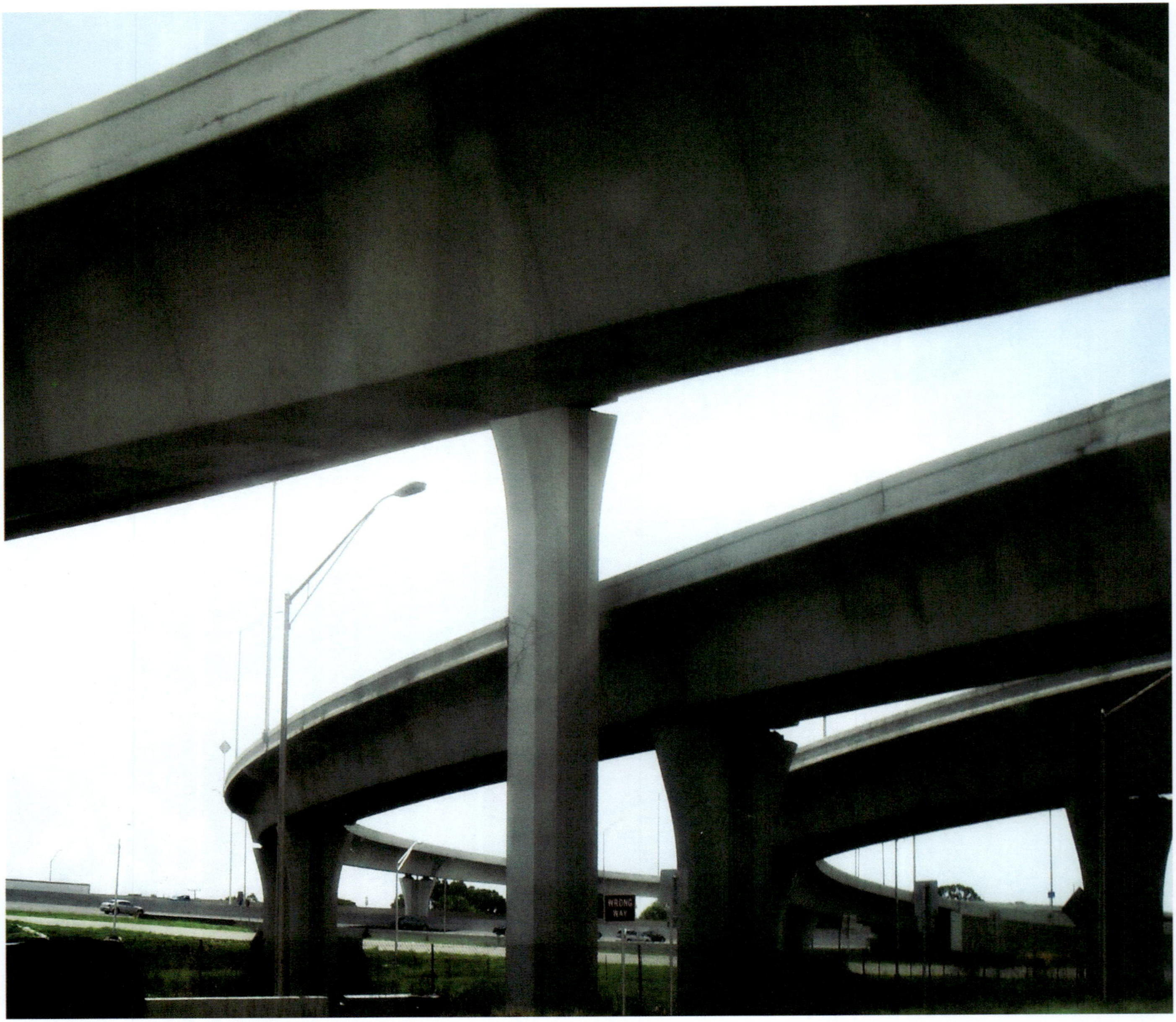

See harmony of line and purpose.

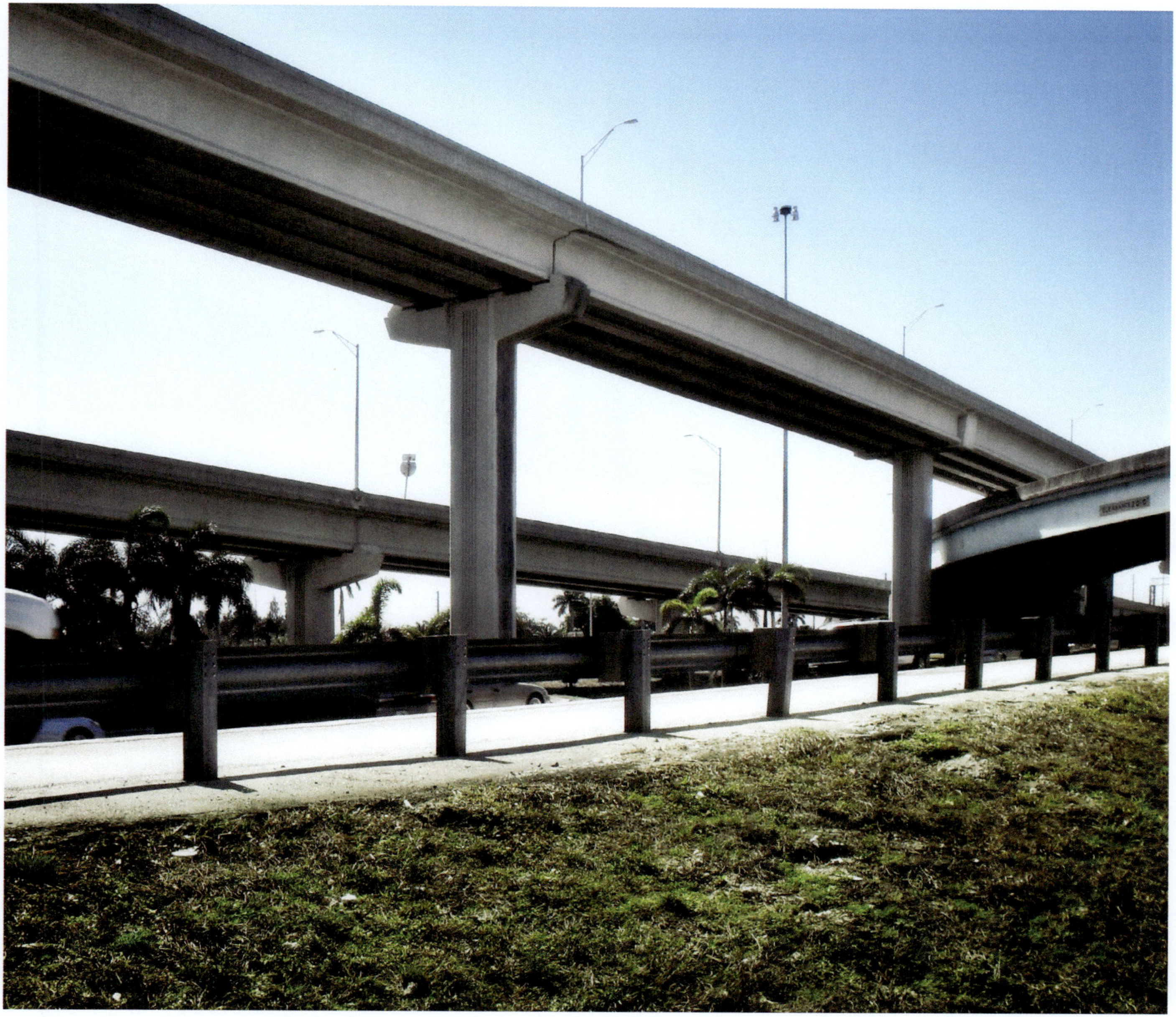

Roads and ramps meld with bridges to transport you across the landscape.

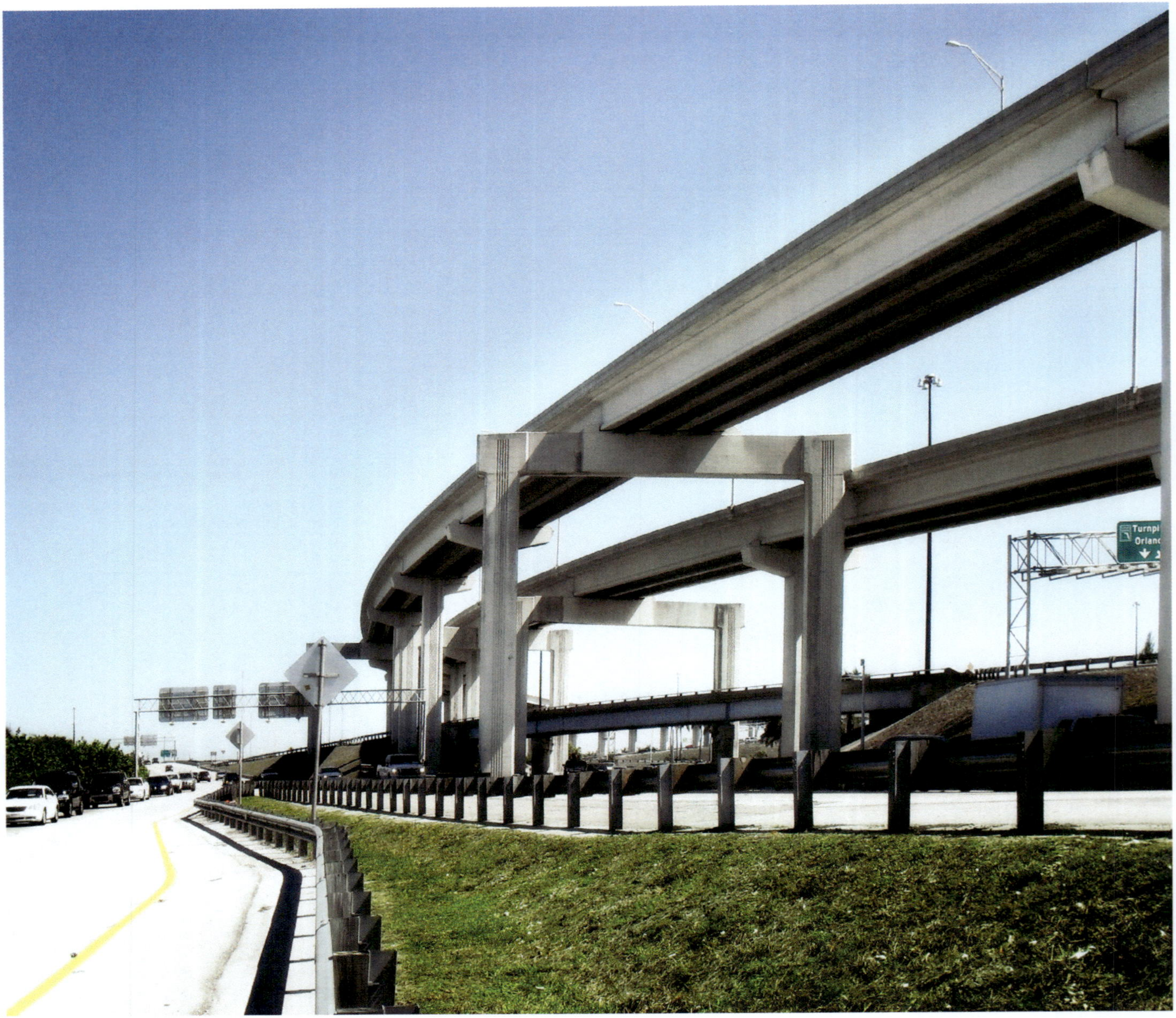

Everywhere you see the concrete praying mantis of the road.

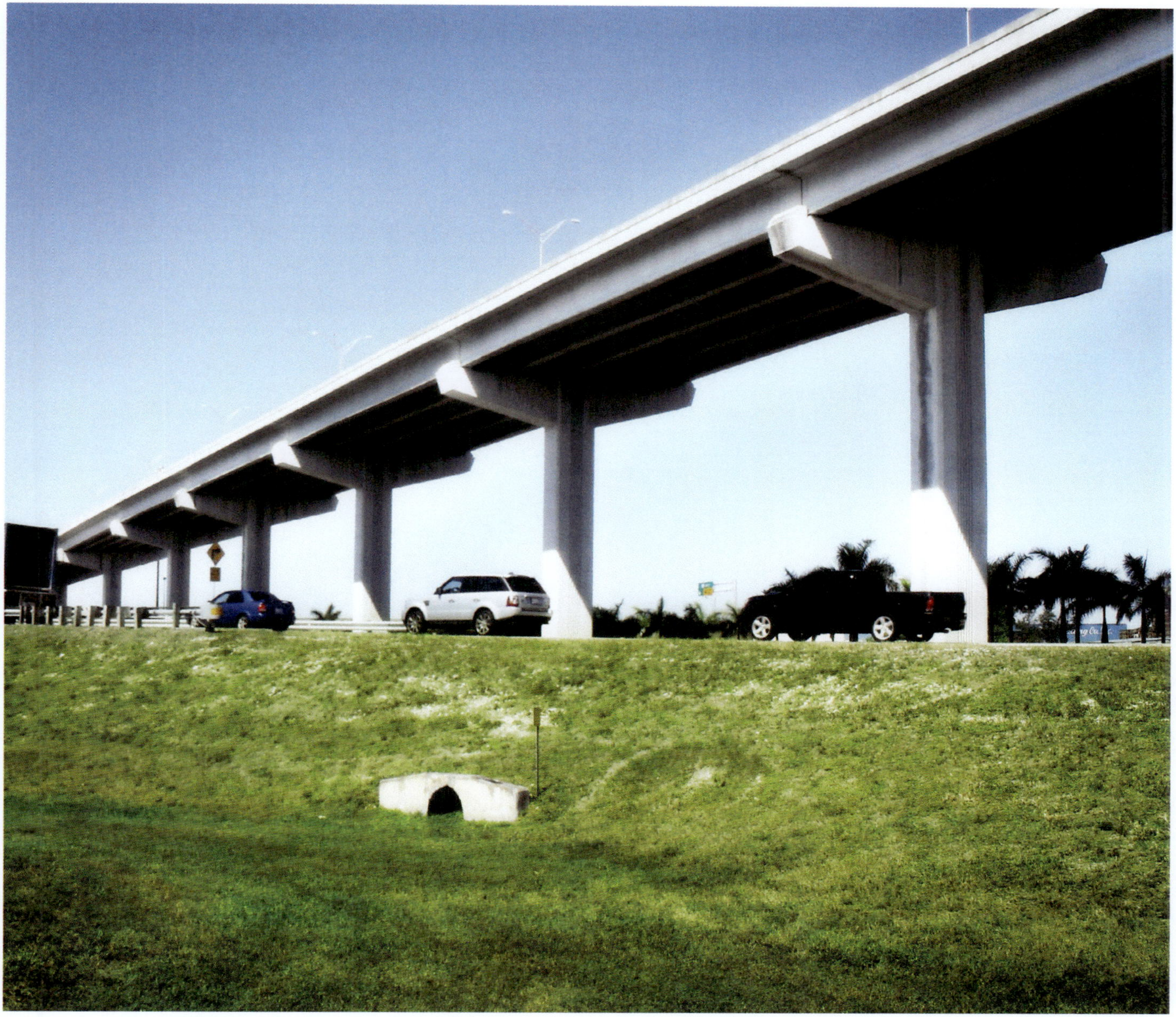

The Bridge's strength holds beauty in simplicity while complex in purpose.

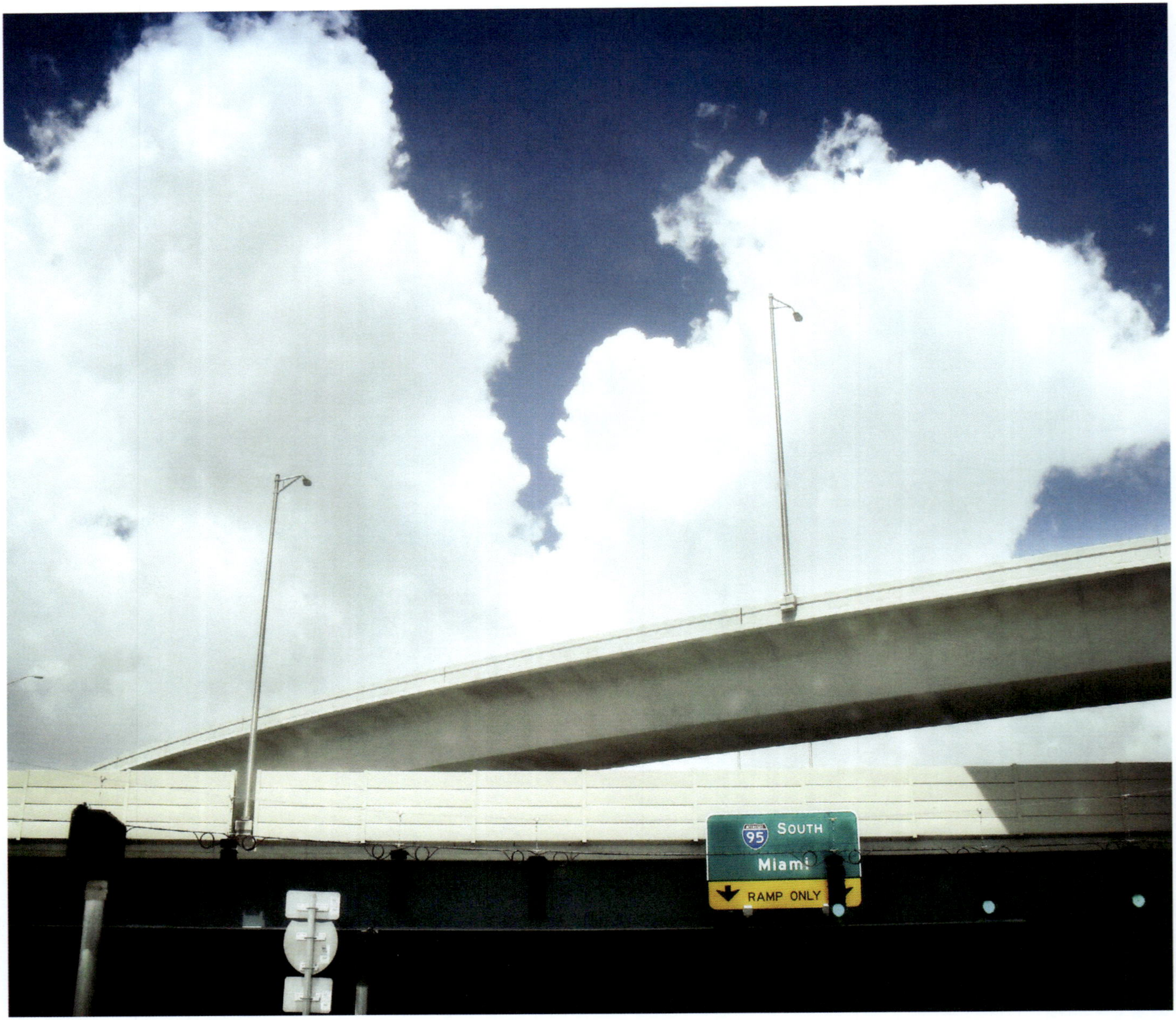

The destination is within reach.

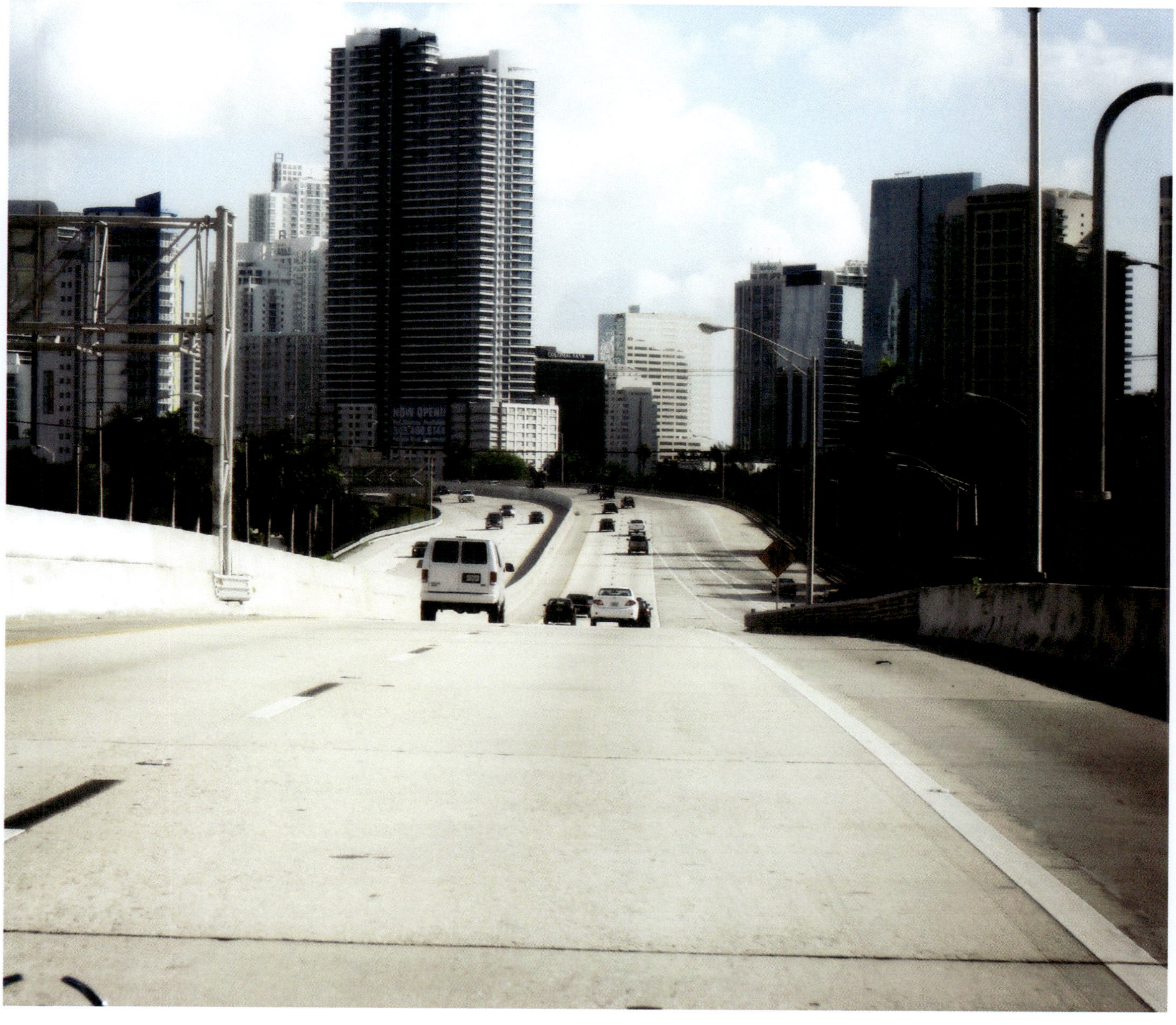

The journey has ended for the day, tomorrow is waiting.

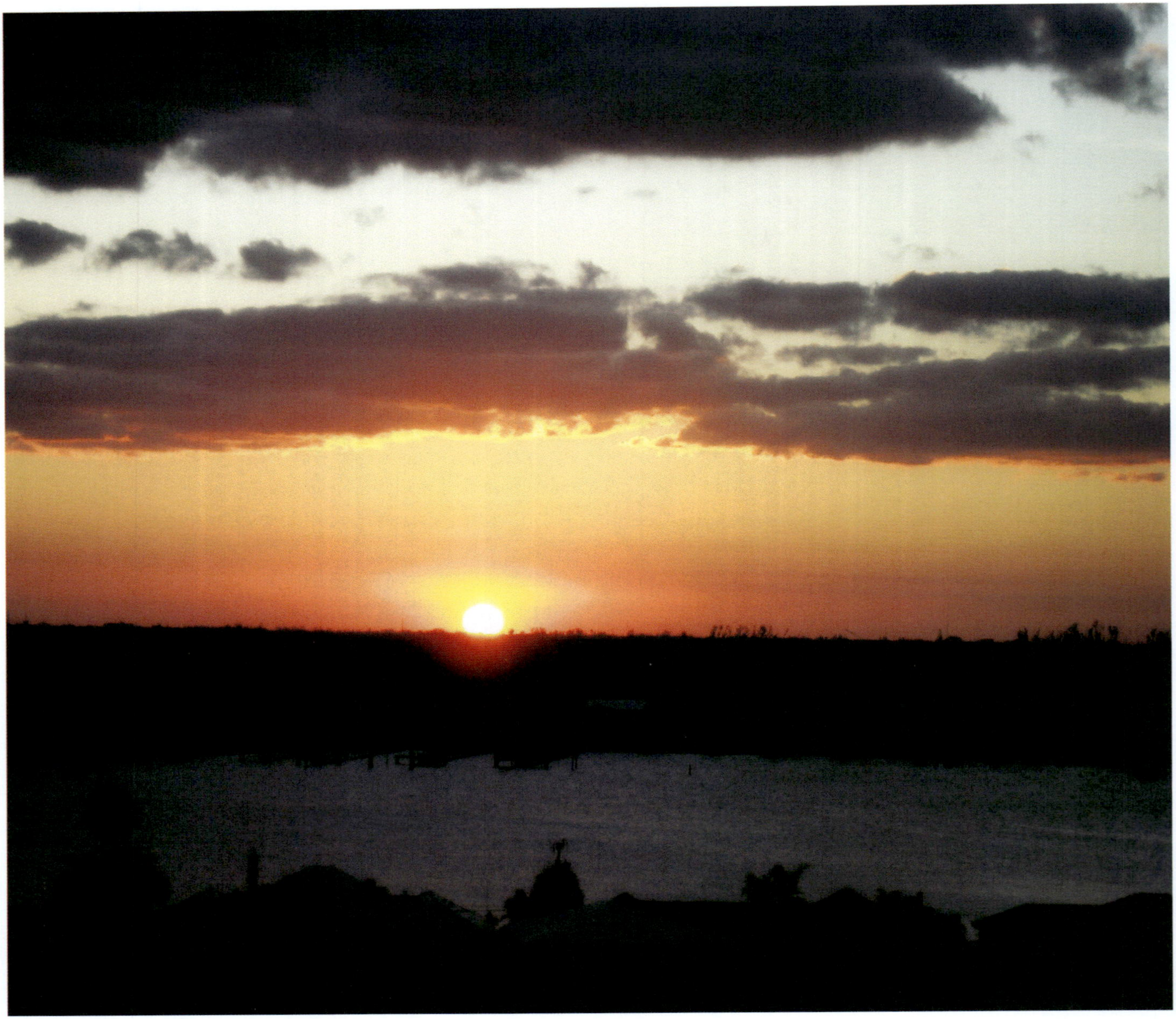

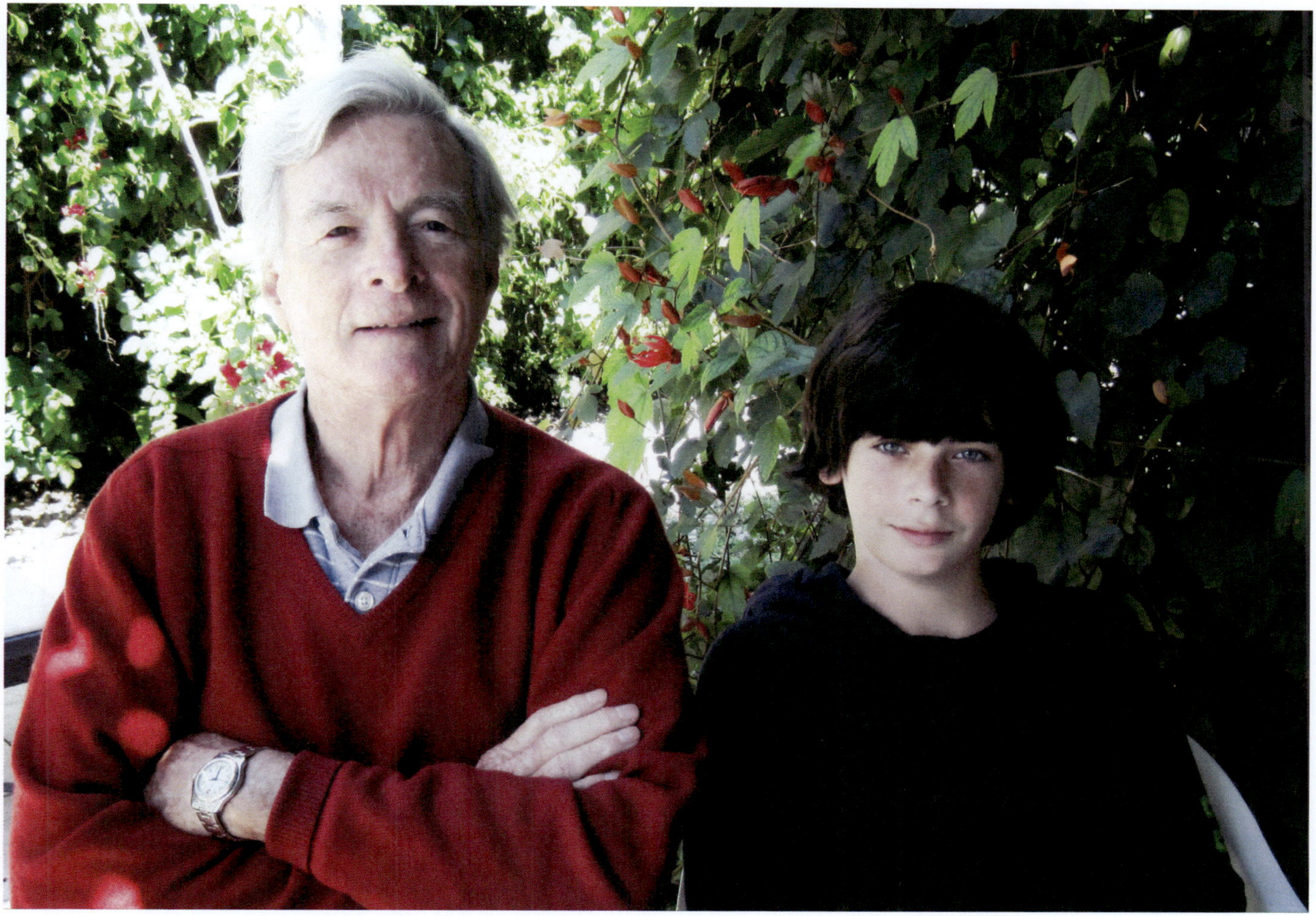

About the Authors Harry + Alex

Harry McQuire has lived and admired South Florida since 1983. He was a student of geology at the University of Colorado and an officer in the air force. In business, he has owned and operated a construction company in the Caribbean, and a brokerage and merger and acquisition firm. He has had a lifelong interest in bridges.

Alex McQuire is an eighth grade student at Palmetto Middle School. His interest in bridges and I-95 started at a very early age and continues. It was during our trips that he pointed out the engineering beauty and huge scope of what the network of bridges and roads entail. He studies science and wants to become a pilot or doctor in the future.

American Red Cross

Helps the community to prevent, prepare and respond to emergencies.

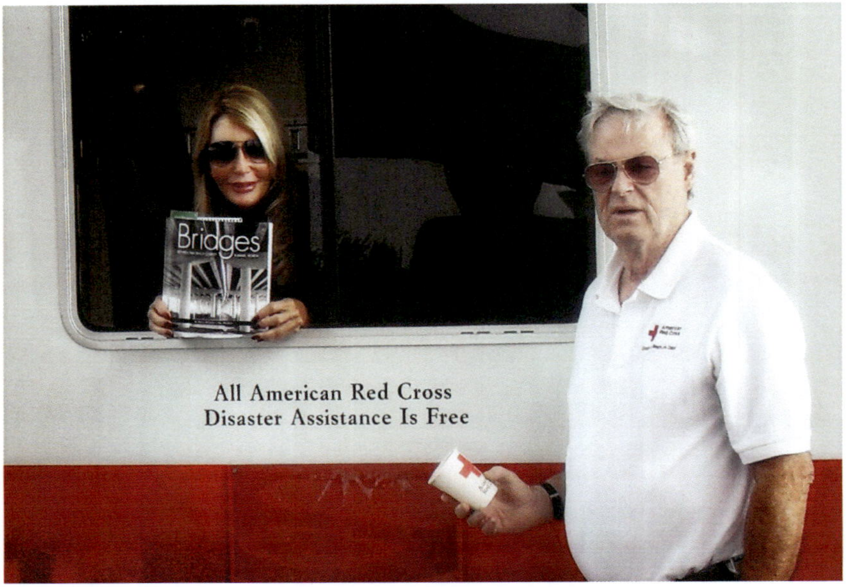

Book design and layout: Amanda McLaughlin
Photo editing and layout: Lois Rothberg

Printed in the United States

Printed in Dunstable, United Kingdom

91222426R00038